Best Value
in Construction

Edited by

John Kelly, Roy Morledge
and Sara Wilkinson

Published in conjunction with

RICS **FOUNDATION**

Blackwell
Science

Blackwell Science Ltd, a Blackwell Publishing company
Editorial Offices:
Blackwell Science Ltd, 9600 Garsington Road, Oxford OX4 2DQ, UK
Tel: +44 (0) 1865 776868
Blackwell Publishing Professional, 2121 State Avenue, Ames, Iowa, 50014-8300, USA
Tel: +1 515 292 0140
Blackwell Science Asia Pty, 550 Swanston Street, Carlton, Victoria 3053, Australia
Tel: +61 (0)3 8359 1011

First published 2002 by Blackwell Science Ltd

2   2006

ISBN-10: 0-632-05611-8
ISBN-13: 978-0-632-05611-8

Library of Congress Cataloging-in-Publication Data is available

A catalogue record for this title is available from the British Library

Set in 10/12.5pt Palatino
by DP Photosetting, Aylesbury, Bucks
Printed and bound in Singapore
by C.O.S. Printers Pte Ltd

The publisher's policy is to use permanent paper from mills that operate a sustainable forestry
policy, and which has been manufactured from pulp processed using acid-free and elementary
chlorine-free practices. Furthermore, the publisher ensures that the text paper and cover board
used have met acceptable environmental accreditation standards.

For further information on
Blackwell Publishing, visit our website:
www.blackwellpublishing.com/construction

Contents

Foreword

It is always encouraging when a profession is looking for continuous improvement and sustains this attitude over a number of years. Identifying new directions is part of this process. This book follows on from the very successful previous volume (*Quantity Surveying Techniques: New Directions*) which I had the honour of editing. It uses the same format and addresses many of the key issues facing construction professionals today including the techniques by which their advice can be enhanced. I believe it will make a major contribution to the advancement of the methods by which construction professionals provide a service to their clients.

Peter Brandon

Preface

In 1990 Blackwell Science published *Quantity Surveying Techniques: New Directions* in association with the Royal Institution of Chartered Surveyors. This represented a distillation of knowledge of the techniques used in the quantity surveying profession at that time.

Time has moved on and it now seems appropriate to re-visit this, to assess what approaches we should be adopting to seek to achieve best value for the client. If there is one factor that characterises the change that has taken place between now and then, it is the focus that is now being placed on the needs of the client and how to interpret these in the most effective fashion. No longer do we assume that the aim is simply to keep costs under control: we now seek to take a far wider view of what we are aiming to achieve.

This book from the RICS Foundation seeks to look at the entire life cycle of buildings, from initial inception, right through to their use and development, and analyses how we can provide best value to the client through the effective application of leading edge techniques and processes.

Buildings play a fundamental role in all our lives and, properly managed, can enhance the quality of life for all of us. This book provides the essential tools to deliver this.

Stephen Brown
Director of Research
RICS Foundation

Biographies

Editor Biographies

John Kelly, BSc, MPhil, MRICS   John is the Morrison Professor of Construction Innovation at Glasgow Caledonian University. Having trained and practised as a quantity surveyor before becoming an academic, he has worked in the field of value management for 20 years, the majority of that time with Professor Steven Male. Together they have published widely, conducted considerable funded research and undertaken value-related consultancy across a wide spectrum of clients and projects. John's major interest lies in the investigation of value strategies at the inception of projects and includes partnering and project briefing.

Roy Morledge, MSc, FRICS   Roy is Professor of Construction Procurement at The Nottingham Trent University. Roy joined the university after 14 years in contracting, latterly as chief surveyor for a regional construction company. He is managing editor of the international *Journal of Construction Procurement* and has published widely through professional and academic journals. He is currently Director of the Graduate School in the Faculty of Construction and the Environment and undertakes procurement-related consultancy work mainly for national organisations on behalf of the University.

Sara J. Wilkinson, BSc, MPhil, FRICS   Sara is a Principal Lecturer and Course Leader in Building Surveying at Sheffield Hallam University. She is responsible for encouraging and developing building surveying research at Sheffield Hallam. Her MPhil research examined the conceptual understanding of sustainable building in the construction industry. Sara was awarded the Jones Lang Lasalle Education Trust Scholarship to research international refurbishment practice in office property. She is a member of the RICS Research Foundation Advisory Panel, and a referee for the RICS COBRA conference. She is a reviewer for the *International Journal of Quality and Reliability Management*. Sara has completed a number of funded research projects and made numerous presentations at national and international research conferences. A number of her research projects have involved international comparisons of surveying and construction practices. Her research interests relate mostly to commercial property and a number of issues such as levels of specification, refurbishment practice, obsolescence and energy issues. She has recently published an RICS research paper entitled *Barriers to Energy Efficiency in the Private Rented Sector*. Currently she has a Teaching Fellowship at Sheffield Hallam University to develop best practice in undergraduate dissertation. Sara's teaching materials are used as examples of best practice in linking research and teaching and available on the FDTL website.

Author Biographies

Donna Duerk, BArch, BA Psych, MArch AS   Donna has been a professor at California Polytechnic State University at San Luis Obispo, California, for 20 years. She teaches architectural programming, universal design, and design studios. She was registered as an architect in Texas in 1973 and has been a member of the Environment Design Research Association since 1972. Her undergraduate degrees were from North Carolina State University and she holds a Masters in Architecture in Advanced Studies from Massachusetts Institute of Technology. Her text, *Architectural Programming: Information Management for Design*, was published in 1993.

David Eaton, PhD, MSc, BSc, ARICS   David is a senior lecturer in the School of Construction and Property Management at the University of Salford. As a quantity surveyor he gained industrial experience with the Property Services Agency of the Department of the Environment of the UK Government, and with Gleeds Chartered Surveyors. As an academic he has completed research and publications associated with quality management, lean construction, zero defects, benchmarking, process modelling and professional culture and ethics. Current research centres on sources of competitive advantage, fuzzy systems, chaos and complexity applications within the construction industry with particular attention to PFI and PPP.

Norman Fisher, PhD   Norman is Professor of Project Management in the School of Construction Management and Engineering at The University of Reading, UK. Between 1992 and 1997 he was the BAA Professor of Project Management. He is also director of Advanced Construction Technology Reading (ACT), a project management process research group. His current research interests include international business and project strategy, marketing processes, project management – design and project process management systems, visualisation and electronic proto-typing/digital mock-up technology, advanced manufacturing systems for construction, automated/lean construction, knowledge based engineering, construction as a manufacturing process, the indentification of user needs for developing design briefs and the wider design process. He is author of a number of publications, including refereed papers, book chapters and four books, with another in progress.

Chris Fortune, BSc, MSc, PhD, FRICS, MCIOB   Chris is a former RICS teaching fellow who gained his bachelor's degree in quantity surveying from the former Bristol Polytechnic, his master's degree from the University of Salford for research into professional competencies in quantity surveying and his doctorate from Heriot-Watt University for research into the building project price forecasting processes. He has held a university research fellowship and he has undertaken a number of RICS Education Trust research projects. Chris is a member of the Association of Researchers in Construction Management (ARCOM) and currently serves as its secretary. Chris is Senior Lecturer in the Department of Building Engineering and Surveying at Heriot-Watt University.

Margaret Greenwood, BSc(Hons), DMS, MSc   Margaret is Senior Lecturer at the Bristol Business School at the University of the West of England. Prior to this appointment, her 20-plus years of working experience and consultancy activities were mainly in construction with a focus on project and risk management. She has degrees in civil engineering, management studies and construction management. Her academic activities centre on lecturing and research in business systems and their management; this includes operations, people, projects and risk. Margaret is currently the Chairman of the Association for Project Management.

John Hinks, BSc(Hons), MA, PhD   John is Director of the Centre for Advanced Built Environment Research (CABER), the virtual research organisation involving the Building Research Establishment, Glasgow Caledonian University and the University of Strathclyde. His present Facilities Management research interests address the strategic nature and assessment of the FM management process. He co-authored the RICS report *Perceptions and Strategic Priorities of Chartered Surveyors* with John Kelly and Gavin McDougall. John has been guest editor of the international journal *Facilities* and is a member of the BIFM Research Committee.

Keith Jones, BSc(Hons), PhD   Keith Jones is a Principal Lecturer in the School of Land and Construction Management at the University of Greenwich and leader of the Sustainable Buildings Research Group. He has been actively involved in building maintenance consultancy and research since the mid-1980s, developing a number of computerised maintenance management systems and authoring numerous academic papers on the subject. He is currently joint coordinator of the CIB W70 International Research Task Group on building maintenance and property management.

Steven Male, BSc, MSc, PhD   Steven is Balfour Beatty Professor of Building Engineering and Construction Management in the School of Civil Engineering at the University of Leeds. Steven has worked extensively with client and construction industry supply chain members in the area of value management. Together with John Kelly he has jointly researched, developed, benchmarked and implemented a value management methodology for the UK construction industry.

Christine Pasquire, BSc, PhD, FRICS, MACostE   Christine is a quantity surveyor and lecturer in the Department of Civil and Building Engineering at Loughborough University where she teaches Construction Economics. She is the Director of the MSc programme in Mechanical & Electrical Project Management and manages several UK Research Council and industry-funded projects investigating aspects of M&E Project Management including whole-life costing, evaluating client benefit, preassembly and early costing.

Jon Robinson, BArch, BSc, MBldg, MLandEc, FRICS, FAPI, FRAIA, FAIB, FAIQS   Jon was appointed to the Chair of Building in the Faculty of Architecture Building and Planning at the University of Melbourne during 1996. Prior to this appointment, he was for many years a senior consultant with AT Cocks Consulting, an independent firm of property advisers and management consultants. His expertise is in accommodation studies, property investment analysis, life-cycle costing, valuation and appraisal and development feasibility studies. In addition he has qualifications in architecture and quantity surveying. Jon has written *Property Valuation and Investment Analysis: a Cash Flow Approach* as well as numerous papers in the field of property including applications in facilities planning and management. He has recently presented papers at international conferences in Helsinki, Amsterdam, Dublin, Reading, Leicester and Singapore. He serves on the Editorial Boards of the *Journal of Property Valuation and Investment* (United Kingdom), *Property Management* (UK) and *Real Estate Valuation and Investment* (Lithuania).

Nigel J. Smith, PhD, CEng, FICE, MAPM   Nigel is Professor of Construction Project Management in the School of Civil Engineering, University of Leeds. Nigel has conducted extensive research into aspects of construction risk and project management and has published widely, particularly in the management of risk in privately financed infrastructure projects.

Lisa Swaffield, BSc(Hons), PhD   Lisa is a Research Associate in the Department of Civil and Building Engineering, Loughborough University. She gained her PhD at Loughborough, studying early costing of M&E services and has ongoing research interests in the briefing process and front-end costing activities including whole-life costing.

Danny Shiem-Shin Then, BSc, PhD   Danny is Associate Professor at the Hong Kong Polytechnic University. Danny's research and consultancy interests lie primarily in the area of strategic space management. He is co-author of the textbook *Facilities Management and the Business of Space*.

David Weight, BSc ARICS   David is the cost data manager for Currie and Brown International and is responsible for the development of their 'Live Options' building project price forecasting tool. David worked as an architectural technician and as an engineering technician before reading for his bachelor's degree in quantity surveying as a mature student at the former Bristol Polytechnic. He has worked for differing types of organisations within the construction industry and qualified as a member of the Royal Institution of Chartered Surveyors in 1996. David was actively involved with the development of spreadsheet applications for energy-efficient cost planning in the mid-1980s which was published by the Society of Chief Quantity Surveyors and led him to co-author a text on feasibility studies in construction with Stephen Gruneberg which was first published in 1990.

1 Best Value in Construction

John Kelly, Roy Morledge and Sara Wilkinson

1.1 Introduction

This book is aimed at construction clients and construction practitioners seeking advice and debate on issues central to the provision of Best Value in construction. This is achieved through commentaries on good practice and the collation of recent research work into construction processes. The book is also structured to provide a pivotal reference for undergraduate and graduate students of construction.

This book gives insights into an understanding of the client value criteria and discusses this in the context of clear strategic and project briefing. The book outlines a range of techniques for managing the project from inception to completion and includes an extensive section on the management of the asset for the benefit of the client. Recent research has evaluated techniques for doing business better such as benchmarking and doing business in a more controlled manner using value and risk management, price prediction and whole-life costing. This book also reflects a changing attitude towards the environment in a chapter on the factors to be considered in environmental management. Recent developments in procurement management, supply chain management and the management of the project are recorded, reflecting a change in the more professional approach to construction by constructors. Facilities management, post-occupancy evaluation and maintenance management, whilst always thought of as important by clients, are now considered vital by all members of the design and construction team. This, particularly as the management of the asset, becomes the financial responsibility of consortia including the design and construction team under procurement systems such as the private finance initiative (PFI).

During the latter part of the 1990s a spate of UK construction reports appeared. Latham was published in July 1994, Technology Foresight a year later, the twelve CIB reports during 1996 and 1997 (see CIB, 1996a–g, 1997a–e) and Egan in 1998.

The UK construction industry has also undergone significant change in reaction to PFI, prime contracting and similar public sector procurement initiatives aimed at satisfying recent HM Treasury directives on value for money and best value. In parallel contractors are undertaking more work against a guaranteed maximum price in which an overt understanding of risk and value becomes vital. Strategic partnering programmes are seen by many contractors as the prime method of offering clients the best service against a background of more certain overhead and profit returns. Involvement of the design and construction team with the client at the earliest stage in the project's life cycle is perceived as being conducive to maximising innovation potential.

Over the same period facilities management and sustainability have emerged as areas of growing importance to the construction industry. In the UK a number of construction companies have rebranded themselves as service providers within the facilities management market – a market with a multibillion pound latent worth.

This presents a rich background for construction research of the type illustrated in this volume. The Latham report *Constructing the Team* published in 1994 set the agenda by stating that improvement begins with clients and particularly with government committing itself to being a best-practice client and promoting excellence in design. This can only occur if a clear and relevant brief prompts a responsible approach to design, particularly in the building services engineering field. Further, the endless refining of existing conditions of contract will not solve old adversarial problems and therefore a set of basic principles is required on which modern contracts can be based and in which adjudication is the normal method of dispute resolution. Building users insurance against latent defects is recommended as being compulsory for new commercial industrial and retail building work. The role and duties of the project manager need clearer definition and in government-sponsored projects, project sponsors should have sufficient expertise to fulfil their roles. Latham recommended a joint code of practice for the selection of contractors, which includes commitments to short tender lists, clear tendering procedures and teamwork on site. Clients should evaluate tenders on quality as well as price. A productivity target of 30% real cost reduction by the year 2000 is probably the best remembered recommendation. The Latham report spawned the Construction Industry Board, the creation of 12 working groups and the production of 12 documents during 1996 and 1997, all aimed at improving construction efficiency by responding to the Latham directives.

The Technology Foresight report *Progress Through Partnership,*

Number Two, Construction, published by HMSO in 1995 stated that the vision for construction is one which sustains high levels of productivity, profitability and responsibility. This was seen to be achievable through the production of world-class products and services for markets at home and abroad and through making major and successive productivity improvements within a new innovative culture stimulated jointly by government and the industry.

Technology Foresight identified five engines of change:

(1) The provision of more appropriate education and training to meet the needs of a modern construction industry, incorporating learning networks to foster greater collaboration across industry and supplier boundaries.
(2) The exploitation of the information revolution to aid communication.
(3) Fiscal changes to facilitate a long-term view and a corresponding increase in the volume of construction.
(4) The creation of a culture of innovation based on the belief that sustained profitability flows from innovation in both process and technology.
(5) The key opportunities which will lead directly to lower cost and greater profitability through the use of standard components to produce customised solutions.

The Egan Report *Rethinking Construction*, published by the Department of the Environment, Transport and the Regions in 1998, stated that the UK construction industry is underachieving, has low profitability and invests too little capital in training, research and development. Many of the industry's clients are dissatisfied with its overall performance. There are five key drivers of change which need to set the agenda for the construction industry at large:

(1) Committed leadership
(2) Focus on the customer
(3) Integrated processes and teams
(4) A quality-driven agenda
(5) Commitment to people

Egan further set the following measurable objectives of an annual reduction of 10% in construction cost and construction time, and a reduction of defects in projects by 20% per year. Achievement of these targets is thought only possible if the industry makes radical changes to the processes through which it delivers its products. The industry should create an integrated project process, which is explicit and transparent to all in the industry and its clients, around the four key elements of:

(1) Product development
(2) Product implementation
(3) Partnering the supply chain
(4) Production of components

Egan's view is that competitive tendering must be replaced with long-term relationships based on clear measurement of performance and sustained improvements in quality and efficiency.

Encouraged by these reports and the announcements made by HM Treasury there has been a move away from the procurement of projects based on the lowest price for a given specification to a more integrated team-based approach to achieving best value. In addition, teams are being required by clients not only to understand the value system of the client organisation but also to design and construct a facility in accordance with the client's value system and often to take responsibility for the running of that facility during its operational life. These issues are addressed in the following chapters.

1.2 Building the value case

Chapter 2 describes the components of the client value system in the context of understanding the corporate value, the business value, and the value drivers for construction projects. The chapter introduces the number and complexity of components of the client's value criteria, particularly within multi-headed organisations in which each department is competing for scarce resources within the corporate whole. The chapter reviews recent research at doctorate level that uncovers the paradigms and perspectives of complex client organisations and identifies the existence of a value thread, which extends through the client organisation and into the project. It is vital for the success of the project that this fragile value thread is not broken during the exchange of information and the passing of information through the complex communication networks that typify a construction project. The chapter also describes the existence of the value chain, a much more robust vehicle for the transmission of information within the various supply chains that exist to progress the project to completion. Supply chains are also discussed in detail in Chapter 11. A view is put forward that the breaking of the value thread is threatened by the procurement route adopted. This theme is further developed in Chapter 10.

1.3   Briefing

Chapter 3 defines and describes the briefing process, giving a review of current approaches, noting the most frequent impediments, and discusses the issues associated with good briefing. Having outlined the hazards to good briefing practice, the chapter highlights the importance of recognising the difference between the strategic brief and the project brief. In order to begin briefing it is necessary to recognise the purpose of the briefing documents, the responsibilities being taken by various members of the construction team, the change management regime, the constraints, the drivers and the language to be used to ensure complete understanding by all members of the team whether construction professionals or lay client members. The chapter discusses the approach to briefing through investigation or facilitation, illustrating many of the points through an example. The chapter concludes by stating that the brief is a document that will contain the project mission and goal descriptions from the strategic brief along with the performance specification requirements of the project brief. The full design and construction team should easily understand each of the requirements in terms that answer the client's needs and desires in the appropriate manner to meet the client value criteria. The brief becomes the primary audit document.

1.4   Benchmarking

Chapter 4 continues the theme begun in Chapter 3 in debating a practical approach to improving performance by describing benchmarking as the application of the skill of comparison. The chapter describes the techniques primarily used in the manufacturing sector and discusses the extent that those techniques can be transferred to construction. A useful checklist of factors to consider precedes a description of prerequisites, processes and methodologies. A discussion of the concept of performance improvement leads into the benchmarking case study, which uses the Movement for Innovation (M4I) industry average performance table of metrics comprising the ten key indicators of construction industry performance.

1.5   Value management

Chapter 5 outlines the background to the introduction of value management to UK construction based on many of the methodologies adopted by US manufacturing organisations. The four generic stages in the development of any project, where the definition of the project is the investment of resource for return, are described in detail with reference to the relative impact of action at each of the four stages on the measure of quality that can be achieved in the final project. A brief review of the international benchmarking of value management precedes a description of the activities undertaken characteristically at various stages during the development of a construction project. The derivation of the function logic diagram, considered by many to be the fundamental aspect of value management, is described in detail. The chapter concludes by stating that value management has reached a level of maturity within manufacturing and construction whereby the style and content of the various workshops is reasonably predictable. However, it recognises that considerable further research work is required into methods of measurement of the client's value system in a form which would be suitable for auditing under the public sector requirements of the various value for money initiatives.

1.6   Risk management

Chapter 6 commences by defining uncertainty and risk. Uncertainty is defined as the term used to reflect genuine unknowns that could have positive or negative effects on a project. Risk on the other hand relates to an event, the time and cost consequences of which together with the probability of the event occurring can be estimated. Risk has a negative effect only. In a construction context it is therefore risk that needs to be identified and managed. Chapter 6 describes risk management over the project life cycle from project viability through feasibility and design and construction operation and maintenance. Techniques to be used and the stages to be recognised are highlighted and the chapter concludes with a case study of a Defences Estates prime contract in which all of the risks are described in detail. A feature of the risk management methodologies described in Chapter 6 is that they bear a relationship with value management described in Chapter 5.

1.7   Building project price forecasting

Chapter 7 outlines building project price forecasting in terms of current theory and practice. Through an extensive survey of techniques in use it has been established that many of those described in detail in cost planning textbooks have fallen out of use. The chapter encouragingly reports that newer techniques relying on computer-based modelling are emerging, typically those associated with life-cycle costing and risk analysis. A detailed case study reviews the 'Live Options' software package and its characteristic ability to be integrated into the work of the design and construction team.

1.8   Life-cycle/whole-life costing

Chapter 8 describes the principles and traditional techniques of life-cycle/whole-life costing and illustrates these with reference to PFI-type projects. The difference between life-cycle costing and whole-life costing is described as being that the latter takes account both of expenditure and a revenue stream. The key concepts of life-cycle costing are described with reference to the time value of money, the life of the building and the period of time the investor has an interest in the facility. The barriers to the successful implementation of life-cycle costing are considered in detail. All techniques are illustrated by reference to PFI methodology. A review of the future is given by reference to current funded research projects.

1.9   Environmental management

In Chapter 9 environmental management is defined and discussed in the context of property and construction. The main issues considered are energy use and global warming, natural resources and waste and recycling, pollution and hazardous materials, internal environment and indoor air quality and planning and land use. The issues are discussed in terms of the decisions required to minimise the impact of the property on the environment. The issues are then related to a case study of a property development proposal for a sustainable building in Melbourne, Australia. Some of the broad building configuration and specification parameters required for an environmentally sustainable building are discussed. It is concluded that by adopting a paradigm of environmental design and construction, the property and construction industry can both play a useful part in the improvement of the environment and reduce humankind's environmental impact.

1.10   Procurement strategies

Procurement facilitates the formal configuration and realisation of a project, where a project is defined as the investment of resources for return. In the context of construction, procurement is the outcome of a complex matrix of decision-taking by or on behalf of the construction client. Earlier chapters have highlighted the importance of understanding the client's business case. Chapter 10 addresses the development of a procurement strategy by analysing the client's business case and client need, prioritising objectives, making overt the client's attitude to risk and assessing the ability of the client and the construction industry in general, to resource the project.

Experienced construction clients adopt a procurement approach which has worked previously or is considered suitable, taking into account their prioritised objectives and attitude to risk. Thus a number of different procurement strategies are recognised and formalised. These procurement strategies are described together with a methodology for the selection of an appropriate procurement approach. The dysfunction that can occur between those involved with the development of the business case and those involved with project delivery is discussed in the context of a correctly considered and recorded approach to procurement. Formal and correct procedures will ensure that the construction industry's approach to construction project procurement is routine and a true reflection of the client's business objectives.

1.11   Supply chain management: a construction industry perspective

Chapter 11 explains that supply chain may be described as the procedures and activities which take a product or service from raw components to something of value to a client or customer. Supply chain management is therefore a structured approach to the organisation and running of the supply chain to maximise competitive advantage. This definition implies that the structure or 'map' of the supply chain is understood, that information is capable of being fed up and down the chain, such that the appropriate members of the chain have the correct knowledge at any particular point in time. Additionally, supply chain management anticipates feedback and learning within the chain that leads to continuous improvement. Within the supply chain there is an assumption of logistical efficiency leading to zero waste, zero defects and zero inventory. Although the science of logistics is rooted in manufacturing, the art of logistics is very much the preserve of the construction industry.

Partnering is recognised as a key facet of efficient supply chain management, particularly in the multi-organisational context of construction project activity. However, the cultural barriers of inter-organisational trust are a major deterrent to effective supply chain management unless effectively overcome as in the case of the Defence Estates' Building Down Barriers project, the case study to this chapter.

Partnering within supply chain management is an effective part of the toolkit of a modern construction industry in satisfying clients that they are demonstrably receiving best value.

1.12   The management of a project

The efficient management of a construction project is intrinsically linked to the organisational structures set up for its management. Chapter 12 focusses on the role of the project manager, whether the client, an employee of the client or an external consultant to the client organisation. Efficiently planning and implementing the procurement strategy is described in the context of resource planning, organisational structure and contractual arrangements. Systems and controls are discussed in the context of time, cost and quality. The chapter brings together the essence of an effective project execution plan.

1.13   Facilities management: assessing the strategic value

The complexities of measuring the value of the facilities management contribution to a client's business is discussed in Chapter 13 by critiquing the limitations of current approaches to measuring facilities management performance. A review of the facilities management literature indicates that much of the current emphasis on performance measurement is driven by a general comparability of indicators, which are facilities-orientated rather than business-orientated. To date most of the published indicators use quantitative assessments of distinguishable and measurable dimensions of the facilities or the facilities function. The chapter discusses the significance of facilities management decisions on overall business success and the dangers of being unaware of this significance. This implies that if the business view of facilities management is to develop then this may have to be led by a corresponding broadening of what facilitator managers view as their consultancy service to clients.

The chapter emphasises that if the performance assessment of facilities management is to expose the usefulness of facilities management to the business, then there needs to be a more overt recognition of the relevance of the business environment to the role of facilities management. The question of the variability in organisational capability to properly exploit the potential of facilities management has rarely been discussed. This chapter addresses this issue by looking at five example scenarios.

The benefit of facilities management, in terms of business value, may reside in its usefulness when applied strategically for competitive advantage. Therefore, the key performance indicators should be directed towards business outcomes, for example agility, flexibility, business continuity and/or transition management. However, the nearer facilities management comes to achieving strategic usefulness, the more complex the method of measurement, and herein lies the dilemma.

1.14   Post-occupancy evaluation

The growth of the facilities management market over the last two decades has given credence to the need for organisations to strategically plan their facilities' or workplace requirements. Chapter 14 quotes a definition of post-occupancy evaluation as being the process of evaluating buildings, occupied for some time, in a systematic and rigorous manner. Post-occupancy evaluation focuses on the building, its occupants and their needs and thus provides insights into the consequences of past design decisions and resulting building performance. This knowledge forms a sound basis for creating better buildings in the future. The issue is whether post-occupancy evaluation is only used as a recording tool for informing future decision-making or as a tool for recording the status quo and highlighting opportunities for building performance improvements.

Buildings provide the infrastructure for some aspect of social or commercial human activity and, as physical assets, are durable products that require life-cycle management. Ultimately, the effective management of buildings is about the fit between the facility and its users. In the context of a commercial activity it is recognised that buildings have a much longer life than most assets in business. The building's value is represented by its effectiveness as a supporting resource in the overall value chain of an organisation's productive process. Chapter 14 describes the assets of the business environment as being: financial performance, physical performance, functional performance and service quality performance. Post-occupancy evaluation, as a tool, has evolved to measure these with the aim of gaining a better understanding between form, functions, work spaces, tasks, organisational subculture and the working environment.

1.15 Sustainable building maintenance: challenges for property managers

Chapter 15 quotes both the British Standard and Chartered Institute of Building definitions of maintenance and refurbishment which have a common thread in terms of the management of a facility in such a way that it is retained in an acceptable condition conducive with the activity of the occupier. As in the facilities management chapter this implies an understanding of the activity of the occupier and, in a business context, the strategic demands that they will make on the building. In this respect the built asset becomes a strategic resource that has to be managed within the broader context of the organisation's strategic plan. Performance measures of buildings in use are required and are described as financial benchmarks, performance benchmarks and disruption benchmarks. The chapter concludes by emphasising the need for a strategic fit between the organisation and the long-term planning of maintenance and refurbishment.

References

CIB (1996a) *Constructing a Better Image*. Thomas Telford.

CIB (1996b) *Educating the Professional Team*. Thomas Telford.

CIB (1996c) *Liability Law and Latent Defects Insurance*. Thomas Telford.

CIB (1996d) *Tomorrow's Team: Women and Men in Construction*. Thomas Telford.

CIB (1996e) *Selecting Consultants for the Team: Balancing Quality and Price*. Thomas Telford.

CIB (1996f) *Towards a 307. Productivity Improvement in construction*. Thomas Telford.

CIB (1996g) *Training the Team* Thomas Telford.

CIB (1997a) *Briefing the Team*. Thomas Telford.

CIB (1997b) *Code of Practice for the Selection of Subcontractors*. Thomas Telford.

CIB (1997c) *Constructing Success*. Thomas Telford.

CIB (1997d) *Framework for a National Register for Consultants*. Thomas Telford.

CIB (1997e) *Framework for a National Register for Contractors*. Thomas Telford.

Latham, Sir Michael (1994) *Constructing the Team* (the Latham Report). HMSO.

Technology Foresight (1995) The Technology Foresight Report. *Progress through Partnership, Number Two, Construction*. HMSO.

Egan, Sir John (1998) *Rethinking Construction* (the Egan Report). Department of the Environment, Transport and the Regions, HMSO.

2 Building the Business Value Case

Steven Male

2.1 Introduction

The aim of this chapter is to explore issues surrounding the development of value options and opportunities on projects. In order to achieve this, a number of key terms are presented initially that will be used throughout this chapter. The chapter draws together work on the strategic phase of projects and the project value chain (Male & Kelly, 1992); the recognition of a two-stage briefing process (Latham, 1994), and collaborative work with a number of researchers from industry (Bell, 1994; Standing, 1999; Graham, 2001) and from academia (Moussa, 19999; Woodhead, 1999). The chapter also draws on an extensive action research programme of studies by Male and Kelly that has been under way since the early 1990s, encompassing value management and value engineering studies, procurement studies that have involved PFI, prime contracting, partnering and other procurement routes, and, finally, organisational/business development projects.

Bell (1994) discussed in some detail the concept of value. Historically it has been presented within an economic perspective in terms of the ratio of costs to benefits and where the primary mechanism to communicate the impact of value decisions has been in monetary terms. However, other authors have presented value in terms of use qualities, esteem features linked to ownership characteristics; exchange properties; cost characteristics, normally the sum of labour, material and other costs. Value can be looked at from the producer's or consumer's/user's side and it has also been related to functional aspects of use. Value has a utility dimension. Bell (1994) captured the essence of this range of factors when she defined value as the intrinsic property to satisfy. In this context it is linked to individuals or groups of individuals and therefore introduces a further issue – that of complexity. Value is therefore dependent on the complexity of perceptions involved, the context within which judgements about value are made, the number of interfaces that exist between individuals, groups of individuals, or organisational units deciding on best value and also the number of organisations involved in the judgement process.

Borrowing terminology from the discipline of 'systems thinking' (Checkland, 1981), a value system comprises people making judgements about best value and value for money. A value system is a complex, organised whole existing in an environment, and is delineated from other value systems by a boundary. It is structured hierarchically and has a common purpose and objectives that can, at times, be in conflict when judging best value and value for money.

The strategic phase of projects and the strategic development process for projects is set out in Fig. 2.1. Client organisations have a strategic direction, normally worked out and expressed in the form of either a corporate plan, for larger organisations made up of many businesses, or a business plan for smaller, single-entity businesses. It is often possible that due to the hierarchical structure of large corporate organisations a project or projects may start their life at corporate level and are then taken up at individual business level or vice versa. Equally, in larger organisations with more autonomous separate businesses, a project or projects may start life at this level, with minimal if any input from corporate level. For smaller organisations projects will remain within the single-entity business. Much will depend on the size, complexity and strategic importance of the project(s) to the corporate and business levels, the investment

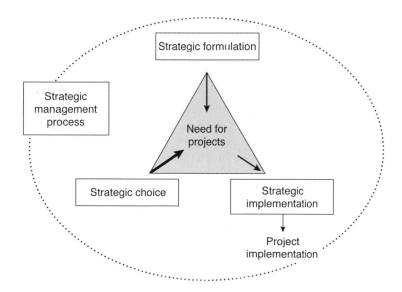

Fig. 2.1   The strategic development of projects.

required and the policy and operating procedures for handling such projects. This sets the parameters for the project development process and the strategic phase of projects.

Recent work by Graham (2001) has indicated clearly that there is a difference between the start of the strategic phase of a project and the start of the project. For privately financed international water infrastructure projects Graham indicated that the gestation of projects during the strategic phase could take many years. Hillebrandt (1984) reported a range of time periods for projects to emerge into the construction industry and Woodhead (1999), in his recent research, indicated that time spans for the projects that he studied varied between six months and three years, depending on the type of client. Often projects will have gone through a variety of changes within the client organisation before they emerge as projects with a clear momentum ready for the industry to commence its work. They will have a history that they bring into the present and that may also influence the future. The strategic phase of the project development process is therefore often messy, fuzzy or ill-defined and can be difficult to pin down in terms of a clear start date for a particular project. The next section looks at client requirements and the client value system as part of the strategic phase of projects.

2.2   Client requirements and the client value system

Large, regular-procuring clients of the construction industry are increasingly pursuing innovative approaches to the way in which their projects are planned, designed and delivered to facilitate their business strategies. They are looking for a structured method to manage their project process within the context of their organisational business strategy, and also work closely with the supply chain to maximise value and achieve continuous improvement in construction performance.

As a result, the construction industry is currently going through major change, much of that driven by regular-procuring clients and also through initiatives such as the Latham Report (1994), the Egan Report (1998), The Movement for Innovation (M4I) and the Construction Best Practice Programmes. There is a clear determination by policy-makers at all levels in the industry that things have to change. Regular-procuring clients are already pressurising the industry for a 'one-stop shop' service, coupled with cost and time certainty in the delivery and they are very vociferous in their demands of the industry to meet their expectations as customers. Newer procurement routes such as the private finance initiative

(PFI) and prime contracting, as well as more integrated, team-based routes such as management contracting, construction management and partnering are attempts at drawing together design and construction interfaces and responsibilities. PFI and prime contracting also incorporate the operational phase. Taking a cross-cultural perspective, an analysis undertaken by Moussa (1999) of the requirements of regular-procuring clients in the European and North American airport industries, indicated that the North American airport clients she had investigated had experienced significant changes with project delivery during the past three to five years. Figure 2.2 sets out the lessons learnt from North American clients, whilst Table 2.1 indicates the stated needs of European airport clients and how their North American counterparts have responded. Clients in the UK construction industry are now implementing many of the solutions adopted by North American clients.

Client requirements for projects, the industry within which they operate, or, for example, within a market sector, such as airports infrastructure development, coupled with the client organisation's culture, history, strategic and tactical management systems and procedures will all combine to create a client 'value system' that will impinge directly on, and influence, the project development and delivery process. The client value system is discussed in more detail below.

2.3   Client value system

The client's value system comprises a number of interacting parts derived from the structure and strategic management process operated by the client organisation itself. The discussion on the client value system will differentiate between corporate and business value, as indicated in Fig. 2.3 and is the subject of the next two sections.

Corporate value

Corporate value is used here to define the value requirements that exist at corporate level within a client organisation that has a diverse organisational structure. This may often reflect, therefore, requirements that stem from across a number of discrete business units or across a number of high-profile corporate projects. At this level, the key requirements will be to align projects with corporate and/or business unit missions and objectives.

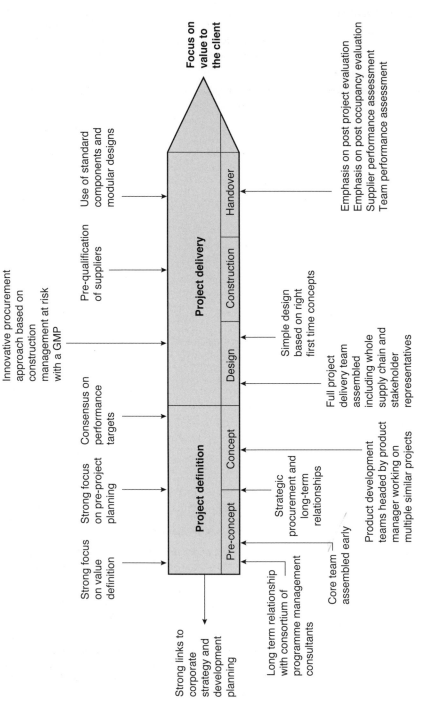

Fig. 2.2   The learning points identified from North American airport clients.

Table 2.1   North American airport clients' solutions to the needs of European airports.

Need for shortening of project time-scale	Innovative procurement routes which allow the involvement of construction firms in the design stage leading to the overlapping of design and construction. Use of packages and fast-track approaches. Greater focus on efficient operations on site.
Need for link between projects and strategy	Strong emphasis on pre-project planning. Strong links between the project development process and development planning. Development managers spending time with business managers to establish project objectives.
Need to address whole life-cycle issues	Emphasis on cost v. value v. return. Involvement of operations, maintenance representatives and other stakeholders as members of the project team.
Need for products to contribute to client's business	Emphasis on pre-project planning, and on the production of a comprehensive brief. Client value criteria communicated to whole supply chain. Understanding of end-users' and business partners' needs.
Need for non-adversarial relations	Creation of teams and focus on team building and teamwork. Dispelling of lack of trust and encouraging mutual respect between all team participants.
Need to build capabilities and capture best practice	Benchmarking of performance indicators. Development of key performance indicators. Development of libraries of information, and project databases. Development of product design standards. Development of progressive improvement initiatives.
Need for long-term supply chain partnering	Use of innovative procurement methods which allow teams to work together over a series of projects.
Need for integration strategies for the whole supply chain	Open book, visible value chain. Development of common objectives. Co-location of team members. Sharing of resources and information technology. Face-to-face communication.
Need for cost reductions	Use of simple designs made up of standard components. Use of modular designs and on-site assembly. Sharing of equipment. Better documentation and design information to minimise excess charges.
Need for early supplier involvement	Greater use of pre-qualification and two-stage tendering procedures which allow for the input of suppliers at an early stage.
Need for simplified processes to eliminate waste and achieve continuous improvements	Process mapping, to make process explicit and reduce variation. Agreed targets for improvement.
Need for improved information transfer	Compatibility of software between supply chain members. Stipulation of software use by clients. Comprehensive project databases and information centres accessible by whole supply chain.

Source: Moussa (1999).

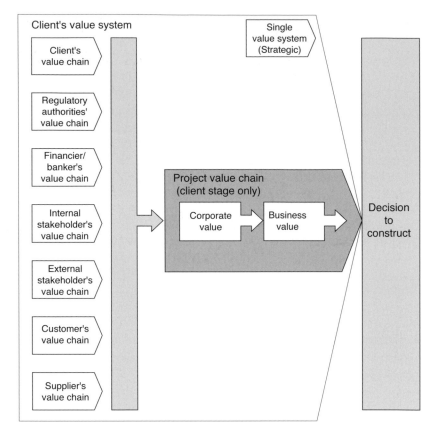

Fig. 2.3 The client value system (strategic phase).
Source: Standing (1999).

Business value

Business value is used here to define the value requirements that exist at business unit level or at the level of a single business entity that does not form part of a larger corporate organisation. Depending on the organisational structure of the client, there will be a requirement to align projects with business unit missions and objectives.

Value drivers for projects

The reason for a project is linked clearly to the ongoing direction of the client's organisation, and has been referred to previously in Fig. 2.1 above. However, as indicated in Fig. 2.4, a project may also be influenced by its relationship to the asset base of the client: for example, if land has to be purchased or sets constraints or

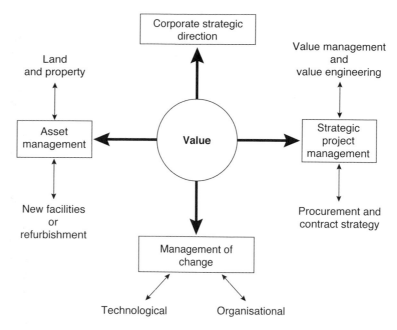

Fig. 2.4　The value context of projects.

opportunities; whether owner occupation is required or lease-back options are being considered; if other projects are competing for resources; whether the project is required as a new facility or as part of a refurbishment programme.

Depending on the client's core 'business' activities, the project(s) may have to adapt to technological and/or organisational change as part of the development process. The value context impacts directly the strategic and subsequently tactical management process for project(s) development. The client value system may have a diffused, as well as profuse, impact and any discussion on the impact of the client value system on projects has to take account of whether the project development process comprises a portfolio of projects or is single-project focused, the subject of the next section.

2.4　The client value system and portfolios of projects and single projects

Corporate or business value can be implemented through a portfolio of projects that comprise a linked network of projects that may be in competition with each other for resources and perhaps competing objectives either at corporate or business unit level. They can also be implemented through a single project. Regardless of whether it is a portfolio of projects or a single project, Bell (1994) coined the term the 'value thread' that must be maintained either through the project network or a single project to ensure value for money is obtained from the formulation and implementation process. When looking from a portfolio perspective, the strategic phase will encompass all competing networked projects. From the portfolio perspective, single-project delivery is a tactical issue. Also, single projects, either delivered within a portfolio or on their own, will have a strategic and tactical phase. Bell (1994) also postulated that there is a value hierarchy within a portfolio perspective, with projects at different stages of their own life cycles. Within an individual project there will also be a hierarchy of value transitions from concept through to use. Thus a hierarchy of values exists between and within a portfolio of projects or within a single project.

Figure 2.5 also mentions the idea of a project value chain, the subject of the next section.

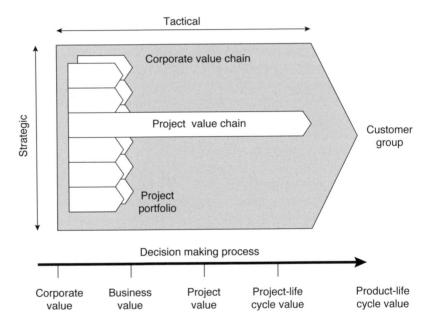

Fig. 2.5   The corporate and project value chain.
Source: Bell (1994).

2.5   The project value chain

Porter (1985) asserts that an organisation comprises a series of internal and external strategically important activities that, when combined, provide it with an advantage over competitors. He termed this the 'value chain'. Therefore, value chain activities are the basic building blocks from which an organisation creates value for the customers of products or services. To date, the concept of the value chain has been applied to an organisation or company's activities but can equally be applied to a project. Male and Kelly (1992) put forward the concept of the 'project as a value chain' within a value management framework for understanding a client organisation's requirements at the strategic and tactical stages of a project. The concept was developed further by Bell (1994) to analyse projects from a large, volume procuring, owner-occupier client perspective involved in project management. The project value chain concept was extended much further by Standing (1999) to look at a project holistically, including the impact of procurement systems.

Standing has argued that the project value chain comprises three distinct major value systems: the client value system, the multi-value system and the user value system. These reflect major value transition points in any project. The project value chain is detailed in Fig. 2.6, which may consist of the portfolio network, where applicable, and/or the individual project with value transitions points of:

- Corporate value within a diverse organisational structure.
- Business value – within a business unit as part of a corporation or as single business entity.
- Feasibility value – where options are generated and a chosen option meets the client value criteria to proceed further.
- Design value – where the designers translate their ideas to deliver value to the client through the design process.
- Construction value – where contractors build the design as briefed and detailed.
- Commissioning value – testing and getting ready for use.
- Operational value – users take over the completed facility and it becomes operational and hopefully functionally fit-for-purpose.

The idea of the project value chain is a useful concept for viewing:

- A network of projects as a series of strategically linked activities that must add value in order for the delivery process to meet the initial strategic requirements of the client.
- Strategically linked activities within a single project delivery process.

This necessitates best value being considered from multiple- or single-project perspectives and sometimes both. Bell (1994), Standing (1999) and Moussa (1999), in particular, have argued that

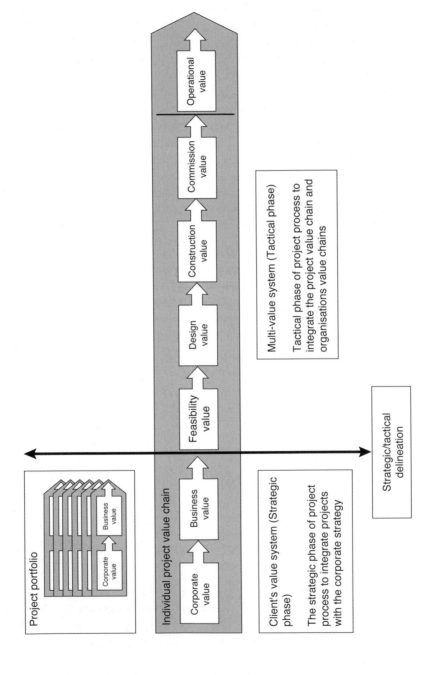

Fig. 2.6   Chracteristics of the strategic and tactical phases of project development.
Source: Bell (1994) modified by Standing (1999).

procurement systems need to take account of the portfolio and single-project value chains.

Procurement and contract strategies are key strategic decisions that may maintain the alignment of the project value chain or create barriers or discontinuities within it. Equally, as a project goes through each of its phases of development, other value transition points occur, for example at feasibility stage, design stage, construction stage, etc. The issue of value and procurement will be discussed later in this chapter.

The decision to construct is an important strategic value point for the client and the project is effectively outsourced (Mitrovic, 1999). The decision to construct is a business commitment that a project is the right solution and capital funding is being made available for further investigation. Other value systems – the multi-value system – then begin to impinge on the project delivery process. The impact of the multi-value system has been investigated by both Moussa (1999) for project portfolios and by Woodhead (1999) for single projects. Both will be discussed in outline.

2.6 Portfolios of project: the value thread and the impact of lean construction thinking

Moussa (1999), focusing initially on airport infrastructure projects, has set out a series of issues that are beginning to be taken up through the use of the newer procurement routes such as prime contracting. Moussa (1999) investigated requirements across a range of regular-procuring European airport clients and also North American airport clients and the potential for lean thinking for the delivery of projects. She defined lean thinking in construction as 'a holistic approach to project delivery focused on maximising value for money to the client through the use of innovative approaches to design, supply, and construction activities'. Implementation of a number of basic building blocks is required to form a coherent whole in order for it to succeed and these are set out in Fig. 2.7.

The application of lean thinking identifies a number of underlying principles:

- Large, regular-procuring clients who have an ongoing programme of capital investment can more easily drive the process. They can provide the continuity of work and the stability required to integrate the supply chain, and practise lean thinking.
- The link between the project portfolio development cycle and the single-project development process makes the formation of long-

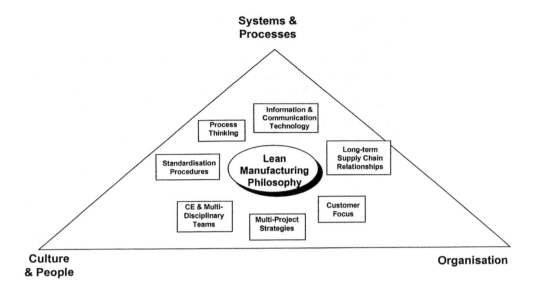

Fig. 2.7   Lean construction and portfolios of projects.
Source: Moussa (1999).

term relationships between clients and the supply chain much easier. This allows a multi-disciplinary approach to projects to occur, since all project participants can be present during the very early stages.

- Embedding single-project processes within a portfolio process provides a more strategic view of single projects, which can provide a process perspective focusing on the repetitive aspects. This permits continuous improvements.
- The presence of a product development stage that commences standard designs prior to the design processes taking place at the single-project level permits standardisation of products and continuous improvement of design solutions, facilitates design for constructability and manufacturability, and allows for the reduction of variability and waste.
- The merging of separate design and construction stages into a single technical project delivery process covering integrated design and production reflects the migration towards concurrent engineering and lean thinking. It also allows for a move towards prefabrication, modularisation and preassembly, which minimise waste and reduce time spent on site.
- The focus on benchmarking and performance measures initiated at the portfolio strategy stage, which are assessed and fed back to the process throughout its life cycle, as well as the specific emphasis on learning and the building of a client knowledge

base, reflects the focus on process thinking and not simply on single projects.

- The explicit focus on knowledge management reflected in the last stage of the project, and the establishment of entities dedicated to documenting process and product information promotes the waste reduction, and continuous improvement ethos of lean thinking.
- The focus on setting up the strategy and performance targets, establishing relationships with the supply chain, and product development reflects the added emphasis on front-end planning.

The project development process proposed by Moussa (1999) addresses six stages at two different levels, the single and the multi-project levels. Between them these stages reflect the changes required to achieve a lean approach to project delivery.

- Stage 1. Creating a multi-project process that will provide direction to the process, define and select a set of productivity and performance measurements.
- Stage 2. Integrating and coordinating the supply chain involved in project development activities.
- Stage 3. Defining projects, investigating their viability and undertaking pre-planning activities.
- Stage 4. Managing the design efforts so they converge to achieve business purposes as effectively and efficiently as possible.
- Stage 5. Managing the delivery process to achieve the standards of performance required.
- Stage 6. Creating and improving the capabilities needed to continuously improve the process over time.

Figure 2.8 sets out these stages sequentially for presentation purposes. The following section explores the single-project perspective.

2.7   Single projects – the impact of paradigms and perspectives on the value thread and multi-value system

Woodhead (1999), building on the work of Billelo (1993), undertook an in-depth analysis of the decision to construct and uncovered an array of different influences that impact the process. He termed these influences paradigms and perspectives. Paradigms are the rules, expectations, values and codes of practice, which are either explicitly or implicitly a part of the professions that play a part within the decision-to-construct process (Fig. 2.9). Perspectives refer

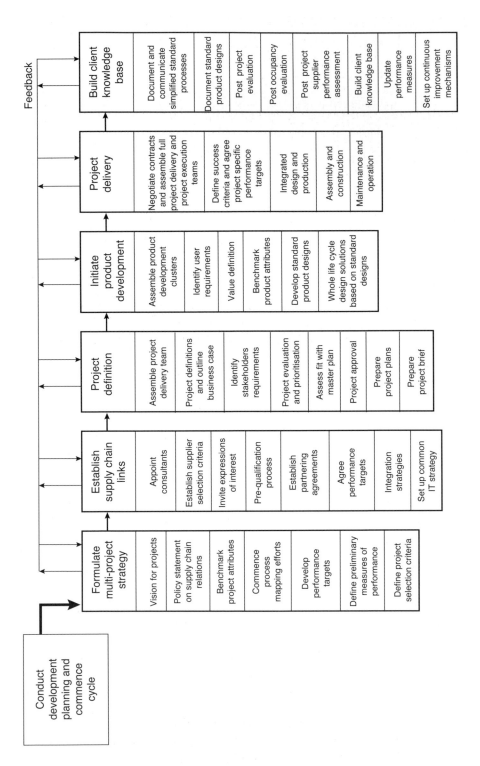

Fig. 2.8 Lean project development process.
Source: Moussa (1999).

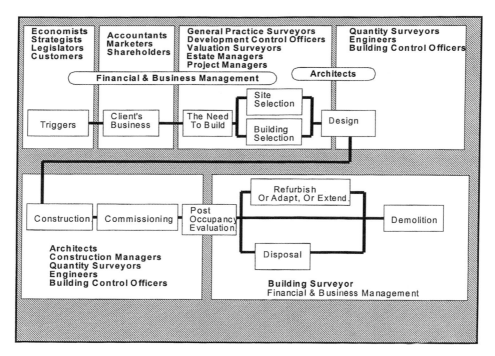

Fig. 2.9 The decision to construct and involvement of the professions.
Source: Woodhead (1999).

to a particular view that can exist within any given paradigm. Woodhead (1999) argued that paradigms and perspectives are often reinforced by the professional chartered institutions, which codify standards and expectations, or through vocational degrees at universities that propagate professional values and paradigms, as well as teach a number of perspectives, or alternative views within each paradigm.

Woodhead (1999) identified the following paradigms and perspectives at play in the decision-to-construct process:

- The capital investment paradigm
- The cost–benefit analysis paradigm
- The financial paradigm
- The strategic paradigm
- The marketing paradigm
- Organisational perspectives
- Management perspectives
- The property development paradigm
- The planning permission paradigm
- The preliminary design paradigm

In addition, he identified that paradigms and perspectives can influence the decision-to-construct process in different ways:

- Those that influence the process of the decision to build
- Those that influence the content of the decision-to-build process
- Those that intrude or are imposed on the decision-to-build process by external agents

During the decision to construct, some paradigms and perspectives will be more dominant than others and conflict can stem from the influence of competing paradigms and perspectives. Thus paradigms and perspectives structure the process of the decision to construct and dictate the content.

Standing (1999) provides a useful diagram that summarises the different value systems – or paradigms and perspectives – that can influence the project delivery process (see Fig. 2.10).

Woodhead (1999) concluded from studying a range of paradigms and perspectives that the decision to construct progresses through a series of stages, each impacted by a series of different decision roles. He identified four categories of decision roles: the formal decision functions of decision approving, taking and shaping, and finally, that of decision influencing. The impact of their roles is identified in Fig. 2.11 as part of the project screening process. Project screening involves, at some stage, the choice of procurement route, which is the subject of the next section.

2.8   Procurement and the project value chain

The choice of procurement route is intrinsically and strategically linked with best value and value for money. Procurement strategy and contract strategy are not tactical choices within projects. They are inherently associated with the management and legal frameworks set up for risk, the delivery of functionality in the design and construction stages and the relationship between time, cost and quality. These value criteria are set out in Fig. 2.12.

The client and design team generate the major cost commitment for most projects. However, they are only responsible for approximately 15% of the client's expenditure, primarily through design team fees. The contractor is responsible for the major element of expenditure, some 85% of the project cost and yet, depending on the procurement route adopted, may be removed from a direct influence on client and design team thinking and their commitment of cost and arriving at best value in the early stages of projects. Hence, when looking at a project from a best value perspective for the client,

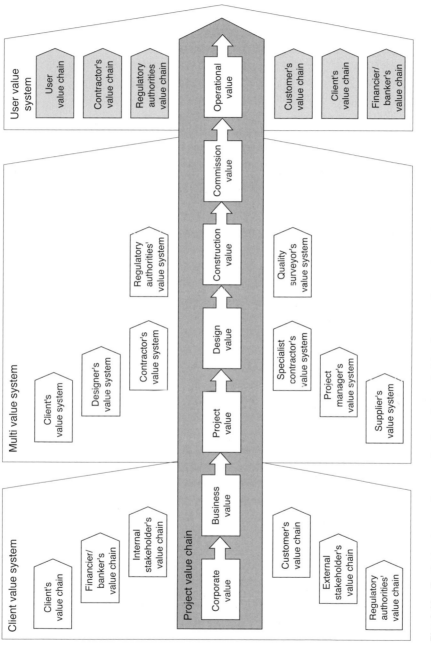

Fig. 2.10 Typical value systems and value chains that impinge on the project value chain.
Source: Standing (1999).

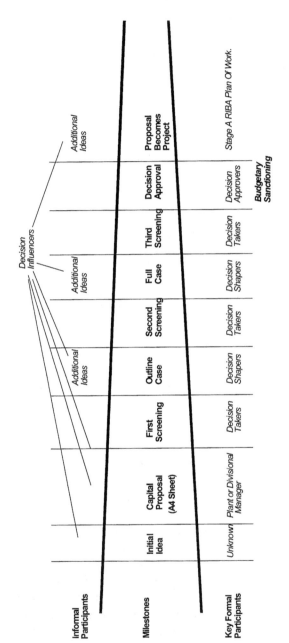

Fig. 2.11 The decision-to-construct process.
Source: Woodhead (1999)

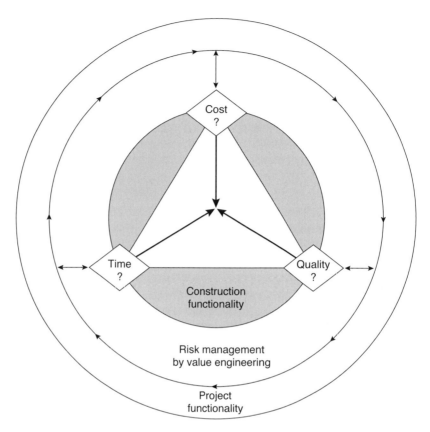

Fig. 2.12   Value criteria – managing time, cost, quality, risk and project functionality.
Source: Standing (1999).

certain procurement routes preclude the contractor's knowledge and expertise from being accessed to the benefit of the project and potentially adding value much earlier in the project process.

Providing the project value chain remains in alignment it is a series of inputs and outputs that create value for the client. As indicated earlier, each value transition should be adding value until the complete project forms an asset for the client's organisation to meet a corporate need. Complexity is added to the project value chain when other value systems impart skills and knowledge into it or create barriers to its effective operation as an integrated system on behalf of the client. One of the most important strategic decisions that impact the project value chain is the choice of procurement route, which can act as either an enhancer or barrier to value creation and improvement.

Figure 2.13 provides a schematic overlay comparing some of the

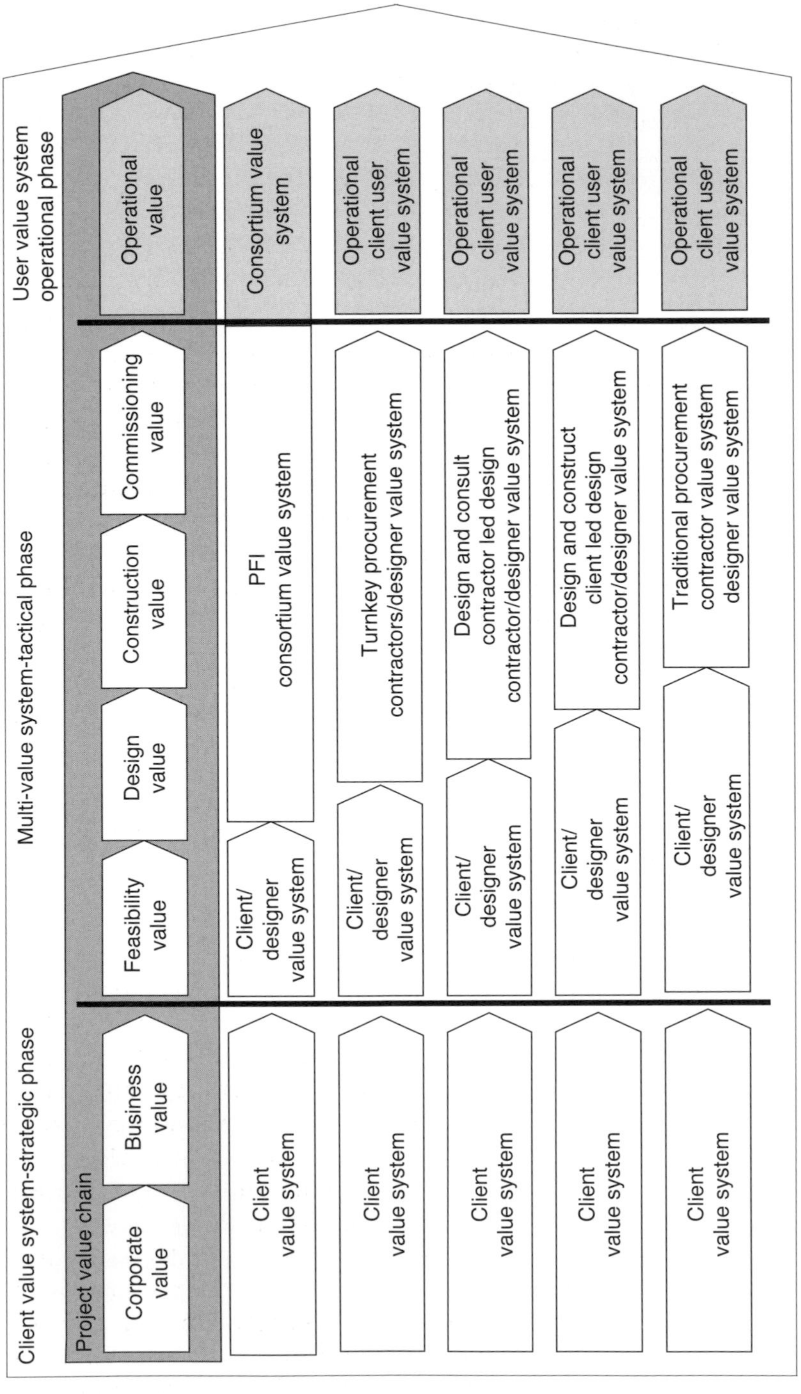

Fig. 2.13 Schematic of procurement systems superimposed over the project value chain.
Source: Standing (1999).

major procurement systems with the project value chain concept. Schematically, those procurement systems at the top of the diagram provide more opportunity to maintain the integrity of the project value chain since an increased number of discrete activities come under one umbrella organisation for single-point delivery. Whilst the project remains within the client value system, value should be maintained internally, although once transferred into the multi-value system through procurement, there is a potential for loss in client value.

In contractor-led procurement systems focused on design and build, including PFI and prime contracting, the contractor offers a one-stop-shop service to a greater or lesser extent and therefore has the potential for increased integration within the project value chain, depending on the method of tender. On the other hand, profession-led design procurement systems involve additional interfaces and provide more opportunities for disruption to the project value chain. Under profession-led systems any change by the contractor must have the client's approval since they are the only party that can sanction change: none have a mechanism permitting a change to occur to the design by the contractor, a topic discussed in outline further below. The management forms of procurement lie some-where in the middle of the schematic and permit increased involvement of construction knowledge earlier in the process but essentially they are profession-led routes with a consequent increase in the number of interfaces.

The use of the PFI procurement system should, in theory, permit minimal disturbance to the project value chain, especially if the client has defined correctly the output specification. Unlike other procurement routes, PFI has a 'double edge' since the consortium has to operate what they have designed and built, potentially for anything up to 25 years. Their focus should be on ensuring continuity and integrity of delivery throughout the process. The PFI procurement system, prior to the introduction of prime contracting, was the only system where the value thread could be maintained into the user value system. Consortia delivering PFI projects have to provide the correct balance of operational expenditure and capital expenditure to increase the level of return on investment. It also provides one of the greatest opportunities to leverage the principles of demand and supply chain management.

Turnkey procurement has similarities with PFI, but unlike the latter, turnkey procurement does not have the additional requirement and liability for operating the facility. The contractual positioning and role of the designers will alter the impact on the project value chain. If the contractor employs the designers in-house then there should be increased alignment in the project value chain.

However, where the designer is independent of the contractor this imposes another value system.

The traditional procurement route, at the bottom of the schematic, is probably the most disruptive to the project value chain since single-stage competitive tendering, if adopted, occurs at the transition point between design value and construction value. Two-stage tendering, overlaid onto the traditional procurement route does, however, have the capability of bringing construction expertise into the project much earlier. Therefore, there is also an interaction between procurement method and choice of tendering strategy in terms of impact on the project value chain.

To summarise, the choice of procurement route is a strategic decision made by the client and/or their advisors that has a fundamental impact on delivering best value and has the capacity to assist or hinder the transfer of value through the project process. The use of value management and value engineering represents methodologies for aligning or realigning the project value chain, the topic of the next section.

2.9   Creating value opportunities: value management (VM) and value engineering (VE)

Put briefly, value management is a team-based activity that is concerned with making explicit the package of whole-life benefits a client is seeking from a project or projects at the appropriate cost. Value engineering, on the other hand, is a subset of value management, is a team-based activity and is concerned with making explicit the package of whole-life benefits that a client is seeking from the technical delivery of the project. Value management has a business focus, is strategic, and is concerned with ensuring that the business and technical projects are defined correctly in the first place, whilst value engineering is tactical and ensures that the correct technical project is delivered during the design and construction phases to meet the business project.

As indicated above, the choice of procurement route is a strategic value decision and can align or disaggregate the project value chain to a greater or lesser extent. VM and VE, depending on whether they are used proactively or reactively, can align or realign the project value chain through a series of value opportunity intervention points as an overlay onto a procurement route or can be built into the route as a formal delivery process. Figure 2.14 sets out the possible value opportunity intervention points highlighted during an international benchmarking study conducted by Male *et al.* (1998). Each value opportunity point has the capacity to bring

The client's 'business project'

The 'technical project'

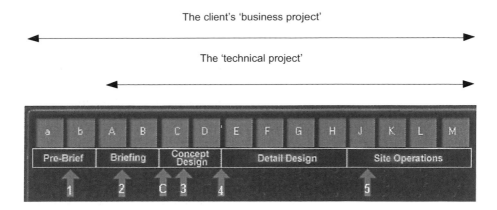

An international VM benchmarking project discovered value intervention opportunities at six points in project development. Each point has a characteristic type of study and workshop. Each is attempting to align or realign the project value chain.

Fig. 2.14   Value opportunity intervention points in the project process.
Source: Male *et al.* (1998). Reprinted with kind permission of Thomas Telford.

together and resolve the competing paradigms and perspectives, decision roles, stakeholder expectations and value transition points that are evident in the project portfolio or single-project delivery process.

Value opportunities can be created at each of the above intervention points, where the right team is assembled at the right time. However, research conducted by Standing (1999) has clearly indicated that certain procurement routes have barriers associated with them in terms of contractor input into the value equation. Given that the contractor spends approximately 85% of the client's expenditure on the project, certain procurement routes, such as the traditional route, prevent contractor value engineering input much earlier into the process. The key issue is the role of the contractor in the project delivery process. The contractor uses intellectual capital to provide innovation and ideas to operate in a competitive market and in that manner gain competitive advantage. The contractor will also manage risk on behalf of the client during the construction process and obtain a reward for that in comparison to a professional fee for a service. Standing (1999) concludes from his research that contractors are not in the market of providing intellectual capital as a service to the industry and contractors seem reticent to cross this boundary and apply a consultancy service into the client's value system. Case studies conducted by Standing (1999) into value management and value engineering by the contractor indicated that innovative ideas generated may only be used once in the project and the possible gain for the contractor can be throughout their own value system. Hence, the contractor must consider the implications of any value engineering change proposal and when and where it could be applied in the context of the marketplace, since a significant competitive advantage could be lost. Within the UK there is no standard value management/value engineering incentive programme to assist earlier input into the project delivery process by contractors, unless the prime contracting or new ACA PCC 2000 Partnering Contract is chosen. Hence, the use of such a programme links directly with the procurement route chosen. The value management incentive programme can operate from pre-contract to commissioning providing the correct framework is adopted in order to protect the intellectual capital of the contractor. Standing (1999) has developed pre- and post-contract value management/value engineering incentive programmes to assist contractor input early into the project delivery process under a range of procurement routes.

The next section draws together the major themes of the chapter to highlight issues that need to be addressed with respect to best value and value for money in the early stages of projects.

2.10 Conclusions

Projects stem from a need or trigger deep within the client organisation. They are an outcome of the client's ongoing strategic management process. The client's value system, an outcome of the strategic management process, has a diffuse and profuse impact on the project development process. Depending on the size of the client organisation and its organisational structure, an ongoing portfolio of projects may be required or a single-project outcome may also result. The project development process has a hierarchical value system that must be taken into account when best value options are generated. Best value options may be generated from questions such as:

- Is a facility required in the first place?
- What is the value context within which project development is taking place and what are the implications for the project development process?
- What is the right project to be delivered, both in business terms and as a technical project? What options are available for the former and the latter?
- What is the right procurement route that reflects the client's value parameters and takes account of either portfolio procurement or single-project procurement?
- What is the right team for delivering this project or projects?

The value thread reflects the ability of the client throughout the project delivery process to communicate their requirements effectively to the various value systems that are comprised of paradigms and perspectives making up the project value chain. At times these may be aligned or in conflict. Value management and value engineering provide a mechanism to align proactively or realign reactively the protagonists within the project value chain.

References

Bell, K. (1994) *The strategic management of projects to enhance value for money for BAA plc*. Unpublished PhD thesis, Heriot-Watt University.

Bilello, J. (1993) *Deciding to build: university organization and design of academic buildings*. Unpublished PhD thesis, University of Maryland.

Checkland, P.B. (1981) Science and the systems movement. In: *Systems Behaviour*, 3rd edn. Open Systems Group. Harper and Row.

Egan, Sir John (1998) *Rethinking Construction* (The Egan Report). Department of the Environment, Transport and the Regions, HMSO.

Graham, M. (2001) *The strategic phase of privately financed infrastructure projects*. Unpublished PhD thesis, University of Leeds.

Hillebrandt, P.M. (1984) *Analysis of the British Construction Industry*. Macmillan.

Latham, Sir Michael (1994) *Constructing the Team* (the Latham Report). HMSO.

Male, S.P. & Kelly, J.R. (1992) Value management as a strategic management tool. In: *Value and the Client*. RICS.

Male, S.P., Kelly, J.R., Fernie, S., Gronqvist, M., Bowles, G. *et al.* (1998) *The Value Management Benchmark: A Good Practice Framework for Clients and Practitioners*. Thomas Telford.

Mitrovic, D. (1999) 'Profitable Partnering in Construction Procurement', in Ogunlana, S., Ed. *Proceedings of the CIB W92 Conference, Harmony and Profit in Construction Procurement*, Chiang Mai, Thailand, January 1999. E & FN Spon.

Moussa, N. (1999) *The application of lean manufacturing concepts to construction. A case study of airports as large, regular-procuring, private clients*. Unpublished PhD thesis, University of Leeds.

Porter, M.E. (1985) *Competitive Advantage, Creating and Sustaining Superior Performance*. Free Press.

Standing, N. (1999) *Value engineering and the contractor*. Unpublished PhD thesis, University of Leeds.

Woodhead, R. (1999) *The influence of paradigms and perspectives on the decision to build undertaken by large experienced clients of the UK construction industry*. Unpublished PhD thesis, University of Leeds.

3 Construction Project Briefing/ Architectural Programming

John Kelly and Donna Duerk

3.1 Introduction

This chapter defines and describes the briefing process and product, gives a review of current approaches to briefing as reflected in the literature, notes the most frequent impediments to good briefing, discusses the issues associated with briefing, examines parties and goals, and summarises a set of best practices illustrated through case study. The chapter uses the terms *briefing*, the term commonly used in the UK, and *architectural programming*, the term commonly used in the USA, interchangeably and without differentiation as they describe the same activity.

3.2 Definition of briefing

Briefing is the process of gathering, analysing, and synthesising information needed in the building process in order to inform decision-making and decision implementation. It implies a focus on the project at two levels – strategic and project briefing. Strategic briefing is the identification of the overall mission or goal of the project discovered before the decision to build. Project briefing involves gathering facts concerning the building project, comprehending the context within which to design for optimum use and aesthetic expression. Importantly, briefing at both levels requires an understanding of all those aspects of the building and its operations that will make it successful.

The brief is used in the design process as a set of evaluation criteria to ensure an optimal solution to the building problem. It is therefore a reference document against which audits can take place at any subsequent stage in the design, construction and use phases of the building process.

The process of briefing involves making a myriad of decisions including setting the scope of the project, prioritising issues for design and choosing from alternative potential solutions. Since no one person has all of the information necessary to make the best decisions, recognising and engaging stakeholders in information sharing and prioritising helps to surface potential problems to be solved before they become insurmountable.

3.3   Current briefing practice

The Banwell Report (1964) succinctly summarised the problems associated with briefing in the following statement:

> 'Many of the difficulties and criticisms of present practices and procedures arise from the fact that those who find it necessary to spend money on construction work seldom spend enough time at the outset on making clear in their own minds exactly what they want or the programme of events required in order to achieve their objective.'

The sentiment was repeated 30 years later in the Latham Report (1994) which concluded that more effort was required to understand clients' needs. Research work undertaken at the University of Salford (Barrett & Stanley, 1999) highlighted the fact that the brief itself was commonly represented as a detailed volume or a few pages recording conversations with the client or a collection of letters in the correspondence file. Rarely is there sufficient material on which to formulate a project to answer exactly client need.

Decisions made during briefing set the tone, the scope and the focus of the design process. Good decision-making strategies are a necessary precursor to the compilation of an optimum project brief. Poor decisions in briefing will certainly compromise the design and ultimately the quality of the building.

3.4   Hazards associated with poor decision-making

The following factors can be associated with poor decisions in general and therefore apply equally to project briefing situations.

- *Intuition* – gathering facts in a random manner and reaching a conclusion, without reference to others, based upon intuitive judgement. This intuitive judgement may be based upon 'rules of thumb' or past experience with no relevance to the current situation.
- *Contextual blindness* – solving the wrong problem through lack of attention to the primary goal.

- *Single context* – identifying in an unstructured manner a single primary goal and rejecting all options that do not satisfy this goal. This could be linked to overconfidence in the correct interpretation of the primary goal.
- *Decision following information overload* – caused by attempting to keep all relevant facts to mind without a systematic method of recording and reviewing.
- *A group is always right* – assuming that a group of people discussing a problem together will always recognise and achieve the right solution even without a structured decision process.
- *A failure to record* – a lack of recording the reason for past decisions in the evolution of the project.
- *A failure to evaluate* – the lack of a frame of reference against which the success of the project will be measured prevents the audit of decisions as they are taken.
- *Feedback blindness* – a reluctance to bring to the client's or group's attention lessons learned from past failure due to ego problems or rationalised hindsight.

3.5   Characteristic hazards in briefing

The literature on briefing (see especially Kelly *et al.*, 1992) provides a rich source of data on some of the hazards commonly observed. These include:

- *Correct representation of the client*. The corporate construction client is a complex entity and it is important to recognise the relative responsibilities of stakeholders and decision-makers. Stakeholders feed information into the briefing process but the final approval of the strategic and tactical brief rests with the decision-makers. It is therefore important to identify and distinguish between these two groups.

- *Is a building the answer to the client's problem?* There is a tendency by the construction industry to assume that because of the client's approach the client has at least correctly identified that a building project of some kind is the correct solution to the problem. Clients are generally assumed to have investigated quite thoroughly the need to build but this may be a dangerous assumption to make and one which is the responsibility of the strategic brief writer.

- *Definition by solution*. There is also a tendency by the construction industry to assume that if the client defines the project by solution, for example 'I need to extend my plant space by 1000 m^2', then the industry will be considered successful if this need is satisfied. However, a strategic briefing exercise might define the project's mission as 'to accommodate new machinery to increase production', which may lead to other solutions.

- *The wish-list syndrome.* It has been found that user groups tend to maximise their 'wish list' in anticipation of being bargained down from this. The problem confronting the brief writer is then to understand the priorities of the user groups such that high-priority needs are not sacrificed for lower-priority wants.

- *Mandatory design guides.* Particularly in the public sector but also amongst large corporate organisations, there are well-defined design guides and standards that tend to be used by the design team without question. The design guides stand in place of the brief for large parts of the project. Outdated or irrelevant design guides can lead to incorrect design decisions.

- *Client change.* A further difficulty facing brief writers is the changes that can occur in the client organisation and the environment during the briefing and design process. A brief can only reflect the needs (and anticipated future needs) of the client at a particular point in time; however, these needs can change during the course of the project in a sudden and unpredictable manner. In contrast, Morris and Hough (1987) demonstrated that in larger projects, history shows that the greater danger is the failure to recognise gradual changes that subtly alter the needs of the client organisation and render the project, as initially conceived, inappropriate. The brief writer should be aware of changes most likely to be implemented in the near future.

- *The less knowledgeable client.* Clients with little experience of the construction industry often do not understand its structure, nor do they have an appreciation of the technicalities of buildings. Problems occur when this type of client is not led carefully through the strategic and tactical briefing process.

- *Lack of iteration in briefing.* Briefing is not a linear process; regular summarising and checking should be a feature of the process.

- *Hidden agendas.* These are a feature of the inception stages of construction projects. Client representatives and the design team members can all withhold their agenda from the group. Exposing hidden agendas by clear presentation and recording of project goals is a function of the brief writer.

The Johari window, Fig. 3.1, modified from the work of Luft and Ingham (1955), illustrates the ideal movement from closed to

Both know	What the client knows
What the design team knows	What neither knows

	What the client knows and the design team need not know
Shared knowledge	
What the design team knows and the client need not know	Managed risk

Fig. 3.1   Knowledge transfer during the briefing process.
Source: Adapted from Luft and Ingham (1955).

sharing knowledge structures. Engendering trust within the group will lead to the sharing of confidential information important to the success of the project.

3.6   Improvements to the briefing process

From the above, a more accurate (in terms of meeting client need) briefing process is addressed in this chapter by considering the following:

- *The parties* – the identification of all of the parties to the process and representation in some way of all stakeholders to the project.
- *The project goals* – understanding the client's intentions from a number of perspectives including time-scales, the amount the client is willing to spend on investigating the decision to build, the underlying organising concept for the project (*a parti*), the perceived stages of the project and the client's value criteria.
- *The issues* – these include the project constraints, project drivers, change management and the responsibilities of those involved in the briefing, design, construction and operations process.
- *The method* – the method and process of briefing.

3.7   The parties involved in the briefing process

As stated above, the corporate construction client is a complex entity with stakeholders and decision-makers involved at differing stages. The parties to the briefing process will comprise stakeholders and decision-makers. Stakeholders include any person with a vested interest in the outcome of a project whether they be the chief executive officer or a nearby neighbour, whereas a decision-maker will be in a position with the authority to make a decision at any point or at a particular point during the process. It is important to identify these two groups and their relative power in the process as it does very little good to get the views and aspirations of people with inappropriate influence. Often it is the person at the top of an organisation chart, the board of directors or a financing institution, that makes the 'go' or 'no go' decision.

Those with responsibility for the briefing process should undertake a test of the relative position of stakeholders and decision-makers at points during the briefing process. A useful test is the ACID test for group membership, described as follows:

A   Authorise. Who has the authority to take decisions at specific points during the briefing process?
C   Consult. Who has to be consulted and in what context? Have those to be consulted a key role to play in the briefing process?
I   Inform. Who has to be informed of any decisions and are these people a formal part of the briefing process?
D   Do. Who is actually going to do each task specified in the briefing documents?

3.8   The project goals

The purpose of the project must be clearly defined and articulated so that all of the parties involved can be aligned in creating the optimal solution to the project's mission. A mission statement is a meta-goal that summarises the direction and intention of the project: it answers the question of why the client is undertaking the project and what outcome is expected. (A full debate on this philosophy can be found in Duerk, 1993.)

Below the meta-goal each design issue will have a goal statement (Pena *et al.*, 1987) that sets out the level of quality that will lead to a satisfactory project. All the intentions that the building is to accomplish should be clearly expressed in mission and goals statements.

Clarifying the mission of the organisation and the purpose of the building project is the most important part of the briefing process in that the building project must achieve a strategic fit with the organisation. Each subsequent decision must forward the progress of the project towards fulfilling the purpose of the organisation. A mission statement inspires a high level of attainment and is rich with the appropriate adjectives and adverbs of implementation. It is more than a sentence describing the number of widgets a factory will produce in a year's time – it describes the aspiration of the client.

Each design issue or problem to be solved should have in the brief a goal statement that assures the client and the designers of the level of quality expected from the final solution. Goals, along with performance requirements, clarify the expectations of the stakeholders and serve as a measure of success for each alternative as it is proposed. Goal statements have sufficient adjectives in them to focus on the attributes of a successful solution.

Implicit in this argument is a view that the briefing process is comprised of two stages: a strategic brief outlining the context and mission, and a project brief with a detailed description of the performance specification of the project. This concept was reviewed by Kelly *et al.*, (1992) and reinforced by the Construction Industry Board (1997). The aim of the strategic brief is to define the project in sufficient detail to be able to take decisions on how to proceed with the project and specifically whether construction is the correct solution to the problem. The objectives of the strategic briefing stage are:

- To analyse all available information and be fully aware of the problem which the project is trying to solve.
- To define the project's function (intention, purpose) as a simple, clear and understandable sentence.
- To make overt the client's value system.

Having taken the decision to build, the project brief is developed as a performance specification to be met by design. The project function and the client's value system from the strategic brief guide the development of the project brief which has the following objectives:

- To understand the users of the proposed building and the way in which the users will interact with the structure, spaces, mechanical & electrical services and technological support.
- To define the spaces by function, size, quality, servicing and IT requirements. This may be expanded into the compilation of room data sheets.
- To understand the arrangement of spaces by adjacency.
- To present a framework against which the design may be checked as it develops.
- To commence the project execution plan document that includes budgetary and cash flow data, key dates and specifically design freeze dates during the development of design.

3.9   The client's value system – *a parti*

An important part of the strategic brief is the *parti* (the organising concept for the whole project). If the clients' priorities show that the image/esteem aspect of the building is the most important part then an organisational/utilitarian concept would be missing the boat. What is needed in this case is a very good function diagram that shows how the building must operate in order to be successful. If functional issues such as circulation/workflow or effective communication/social interaction among the workers are the top priorities then the briefing process would do well to explore the different alternatives to building and site organisation to find the scheme that would best fulfil the top priority needs. This logic extends to other aspects of client preference, organisational strategy and/or political policy.

3.10   Briefing issues

Briefing is an iterative process that progresses from the general to the specific. The first part of the process is to understand the issues surrounding the project and then begin to fill in the blanks. The design is developed in the context of the issues – the client processes and characteristics, the codes and values held by the stakeholders, the values held by the surrounding community, etc. These facts are the information base that the designer will use in making decisions on alternative solutions to the design problem. They are also triggers for a search for deeper information that will help to define a real problem to be solved. In the process of uncovering the existing situation, specific subsets of the design problem will emerge.

The issues surrounding the project can be explored under the following headings:

- *The purpose of the briefing documents.* It is necessary to make explicit the purpose of the strategic brief and the project briefing documents, particularly if these documents are to be used subsequently for audit purposes. In the situation where the briefing documents highlight the characteristics of the success of the project then these characteristics must be clearly set as measurable objectives and incorporated into the project execution plan.

- *Identifying responsibility.* It is important to record the decision-making structures in a client organisation and the way in which these will impact the project. Decision-making structures become more important in situations where a single project sponsor or project manager represents the project team. The limits to the executive power of the project sponsor or the project manager should be clearly defined.

- *Change management.* The client should, during the briefing process, be made aware of the impact of change during the design and construction process. The client should be left in no doubt that making a change at an inappropriate stage will incur financial and time penalties. An important part of the briefing process is to recognise, plan and advertise widely, the design freeze dates. The briefing team should recognise that the majority of client organisations are themselves change-orientated and dynamic, and therefore client organisational change is sure to occur during the design and construction of the project. The brief should include a statement of those areas that may be regarded as fixed, and those that may be regarded as flexible in this context.

- *Recognising constraints.* Constraints may relate to the context of the project in terms of location, the culture, tradition or social aspects relating to the client or the near-neighbours of the project. Location can be a significant constraint for the construction process; for example a congested city centre site, a low bridge, a ferry with a weight limit, all might require a particular design solution. Culture or tradition relates to organisational hierarchy and physical position, corporate identity, etc. The surrounding community members are often overlooked as stakeholders in a building project but their opposition can stop the process.

- *Identifying drivers.* Projects are conceived as a result of a wide variety of drivers. Drivers include new legislation, a requirement for upgrading existing premises, the replacement of existing premises, etc. At the strategic briefing stage it is necessary to identify the strategic drivers. This can often be a difficult area to investigate in organisations bound by politics and hidden agenda. At the project level, drivers are those elements requiring a specific design response such as circulation, image, safety, durability, etc. As the design develops, a designed solution may itself become a design driver to a subsequent stage.

- *Identifying experience.* The design team leader should assess the experience of the client's team in undertaking similar projects and maximise the incorporation of that experience in the evolving design.

- *Language.* The design team should recognise that not all members of the client's team are conversant with the language of drawn information. Therefore, the language most suitable should be investigated at the time of developing the brief. There may be a requirement to express drawn information in the form of 3D models in which case the drawn information from the design team should be configured to best interact with a 3D modelling tool.

- *Legal and contractual issues*. All factors which have a legal bearing on the project should be explored during the briefing stage. The extent to which the client is risk-averse and also requires cost certainty should be made explicit. Projects which by their very nature are innovative may require the expert input from the contractor and the contractor's supply chain at a very early stage in the process, in which case a competitive tender may not rely on lowest price alone.

3.11   To investigate or facilitate?

There are two approaches to the strategic and project briefing processes using the techniques described below, these being investigation and facilitation. In investigation an architectural programmer, an architect or a project manager will compile the brief through a process of literature review, interview, post-occupancy evaluation and existing facilities walk-through. The data thereby gathered will be checked through presentations at team meetings.

In facilitation, a facilitator independent of the client and the design team will guide the whole team through a process of briefing using largely the techniques described below. Using facilitation it is common to have teams comprised of different people for the strategic and project briefing stages.

The advantages and disadvantages of each method are given in Tables 3.1 and 3.2.

3.12   A structured approach to strategic briefing

The process leading to the compilation of the strategic brief involves: firstly, the discovery of facts; secondly, the processing of issues; thirdly, the understanding of the client value system; and finally, the definition of the project mission and/or goals.

The discovery of facts

Data in support of facts for the strategic definition of a project are gathered by:

Table 3.1   Advantages and disdavantages of the investigation method.

Advantages	Disadvantages
A skilled programmer will be able to logically collect data using proven techniques in an efficient manner, minimising client input.	Points raised in later interviews may require revisiting earlier interviewees to clarify issues.
More than one programmer can be used on a project to gather data in the shortest possible time.	Checks need to be made to ensure that interviewees are using language and terms in a common manner.
It is not necessary to identify all stakeholders at the commencement of briefing. If an important stakeholder comes to light during the process they can be interviewed in turn.	It may be necessary to enlist the help of an expert to ensure the right questions are asked.
By interviewing individuals in turn it is always possible to obtain the honest views of those junior in the client organisational hierarchy.	It is difficult to counter the 'wish-list' syndrome, particularly where a forceful stakeholder puts over their requirements as a *fait accompli*.
A confidential interview on a one-to-one basis is likely to uncover hidden agenda.	Data from many interviews will take considerable time to process.
Interviewing will reveal the decision-makers in situations where their identity is not clear.	Validation checks with a team may reveal discrepancies late in the briefing process.

- Literature search – usually in support of a new project type.
- Post-occupancy evaluation – of a similar facility.
- Planning activities based upon visits to the client's sites and observation of the client's organisation.

Projects will have standard facts and project-specific facts. Some facts will be immediately known by the briefing team and others will have to be determined by structured investigation. The former may include site analysis, climate data, relevant planning and building codes, etc. The latter may include historical and future trends, occupancy levels, utility service capacity, site water table, traffic impacts, etc. It is important to recognise that at the strategic briefing stage only sufficient facts are sought to inform the strategic decision of whether to build or not. Once the decision has been taken, a second round of fact discovery will take place in order to inform the project brief.

The discovery of issues

All issues will be determined at the strategic briefing stage. Issues

Table 3.2 Advantages and disadvantages of the facilitation method.

Advantages	Disadvantages
A facilitated team strategic or project briefing exercise will extract all of the information in the shortest time.	A facilitated strategic or project briefing exercise demands a skilled facilitator knowledgeable in briefing.
In a facilitated team exercise any misunderstandings and/or disagreement can be resolved immediately.	A representative client team will contain members from different levels in the client hierarchy, which may stifle contributions from junior members.
A full team will contain all of the experts necessary to feed information into the project and ask appropriate questions of others. This is particularly useful at a project briefing exercise which includes appropriate client representatives and the full design team.	If a key stakeholder is missing from a facilitated team meeting then key information may be omitted. This may be difficult to subsequently incorporate.
The facilitator or other members of the team can challenge 'wish-lists'.	Hidden agendas may be difficult to uncover in a team exercise.
The facilitator will summarise the data contributed by the team at stages during the team exercise, therefore the brief will largely comprise a collation of these conclusions.	It is impractical to undertake a facilitated team meeting without some prior interviewing which requires a measurable investment in terms of client team time.
An intensive, focused, facilitated briefing exercise will encourage good team dynamics and effective team building.	Facilitated team meetings can challenge the authority of the architect or project manager.

are those pieces of data that impact the strategic brief and generally are qualitative in nature. Typical issues are outlined above. One method of gathering issues is by interviewing decision-makers and stakeholders.

Identifying decision-makers

It is important to understand the plans and aspirations of those members of the client's organisation who will make the final decision with regard to the project and thereby understand the context of the views and ambitions of those people who have not got full authority to make a decision. Most often it is the person at the top of the organisation chart, the board of directors or a funding institution that makes the 'go' or 'no go' decision. Early in the process the briefing team should outline the decisions that need to be made and ascertain who has the final word.

Identifying stakeholders

As indicated above, decision-makers are not the only stakeholders on the project. Various users will have a valuable input into the briefing process with regard to information about processes, organisation, and needs and wants. It is important that all types of stakeholders are identified and represented, even if there is no potential controversy surrounding the project. This ensures that vital pieces of information are not left out and that the success criteria of the project are clearly recognised.

Interviewing decision-makers and stakeholders

Throughout the briefing process there are likely to be many meetings with client representatives. Interviews should start with very general questions so that the interviewees are allowed to state their points of view, moving to the more specific questions, which ultimately creates an opportunity to have designers' assumptions scrutinised. Following the interviews, it is useful to have the decision-makers and the stakeholder representatives in the same room so that they can share their information and views on the problems that the project is solving. Often organisations do not take time to undertake this process on their own and this form of project briefing becomes a useful vehicle for the discovery of the interactions of system components within the client organisation. One of the best processes for displaying information as it turns up is to put each idea on a piece of paper on the wall where everyone can see the array. This visual organisation of data and ideas stimulates discussion of values, contentions, needs and wants in a manner that works towards creating a set of priorities to which the whole group can subscribe.

As each conceptual option is developed it is useful to verify with the group that the interpretation of the intentions has been done appropriately. This highlights the language issue discussed earlier.

Clarifying the mission of the organisation and the purpose of the project

This is the most important part of the strategic briefing process and of the strategic briefing document. Following the decision to build each design decision will progress towards fulfilling the mission. This implies that each element of the design will be solved by reference to a goal statement which embodies a common understanding by the client and the designers on the level of quality to expect from the final solution and thereby indicates that the learning process is taking place. In addition, goals along with performance requirements clarify the expectations of the stakeholders and serve as a measure of success for each option as it is proposed.

3.13   The client's value system

The client's value system is at the heart of the goal definition and comprises the following seven elements which may be explored and ranked using the pairs comparison technique:

(1)   *Time* – the time from the present until the occupation of the building.

(2)   *Capital cost (CAPEX)* – all costs associated with the procurement of the building including land purchase, fees, and construction cost and fitting out.

(3)   *Operating cost (OPEX)* – all costs associated with the facilities management of the building which may be limited to maintenance, repairs, utilities, cleaning, insurance, caretaker and security, but may be expanded to include the full operational backup such as catering, IT provision, photocopying, mail handling and other office services.

(4)   *Environment* – the extent to which the building is to be sympathetic to the environment measured by its local and global impact, its embodied energy, the energy consumed through use and other 'green' issues.

(5)   *Exchange or resale* – the monetary values of the building were it to be sold. This concept requires the client to think about the building's future and the time when it may become economically redundant.

(6)   *Aesthetic/esteem* – the extent to which the client wishes to commit resources for an aesthetic statement or portray the esteem of the organisation. Many office towers are built for esteem.

(7)   *Fitness for purpose* – the level to which the building supports the operation of the business in purely utilitarian terms. A very high rating under this heading would imply that a minor part of the budget would be used for art or an architectural statement. Occupation densities are likely to be high. Many call centres are built as being solely fit for purpose.

3.14   The definition of project mission or goals

The purpose of the project must be clearly defined and articulated for all parties involved in the project to be aligned in creating a quality solution. A mission statement is a meta-goal that summarises the direction and intention of the project. It answers the question of why the client is undertaking the project and what outcome is expected. The meta-goal will be subdivided into individual goal statements that define the needs and wants of the client and if appropriately satisfied will lead to a satisfactory project outcome. All the intentions that the project is to accomplish should be clearly expressed in the mission and goals statements.

3.15   The decision to build

A structured approach with respect to the discovery of facts, the processing of issues, the understanding of the client value system and the definition of the project mission and/or goals will lead to the point where the client can make an informed decision to build. This is a fundamental decision to take, as the decision to build will launch complex consortia of organisations whose objective will be the supply and maintenance of built space. The cost consequences of a decision to build are invariably high.

If the strategic brief has correctly represented the mission as a functional statement of need then the decision to build can be taken with confidence.

3.16   An example of strategic project briefing

In this example it is assumed that the National Park Commission wishes to enhance the pleasure of visitors to the National Park while at the same time restraining visitors from doing damage to the natural environment. The Commissioner has decided to build a Visitors Centre with the primary aim of educating visitors.

In this example a facilitated meeting was the technique used to derive the mission statement as is described in the chapter on value management (Chapter 5). The facilitated team exercise involved members of various departments from the Commission and included the key decision-maker who has executive authority to authorise the commencement of the project and a spend of up to $3 million.

The function tree diagram with the mission statement at the top left-hand side is shown in Fig. 3.2. The tree diagram comprises the mission statement, the performance requirements (*parti*), the lower-level performance criteria and concepts to be achieved by the project design.

In order to properly formulate the value criteria of the client a pairs comparison was undertaken using the headings described

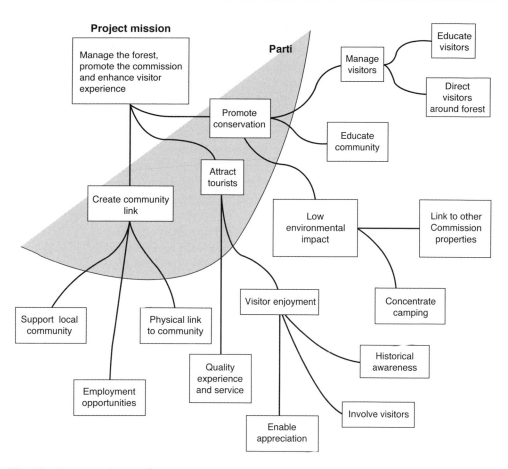

Fig. 3.2   Tree structure for information hierarchy.

above in the section entitled the client's value system and illustrated in Fig. 3.3. This process gives confidence to the 'go' or 'no go' decision. A pairs comparison considers only one criterion against another. In the example below, the question was asked: 'Which is more important to you, A or B?' At the end of the exercise the number of times A occurs gives a measure of the client's value criteria. This is a useful overt measure against which to audit future decisions.

Figure 3.3 demonstrates that the most important aspect of the project is its lack of impact on the environment, aesthetics and function are equally weighted and operating cost is more important than capital cost. The time for construction is not so important and the resale value of no significance whatsoever.

These stages would allow a decision to build to be taken and if the

A. Capital cost

B	B. Operating cost					
A	B	C. Time				
D	B	D	D. Function – fitness for purpose			
E	E	E	D	E. Aesthetic – esteem		
A	B	C	D	E	F. Exchange	
G	G	G	G	G	G	G. Impact on environment
2	3	1	4	4	0	6

Fig 3.3   Pairs comparison of client's value system.

decision was in favour of the construction would inform the development of the design.

3.17   A structured approach to project briefing: description and example

The first stage of a structured approach to project briefing necessitates a review of the mission, the meta-goal of the project and the subsidiary goals relating to the solution of each design issue. Once this and a clear functional analysis is understood, the project brief can be evolved by reference to the development of a space programme and conceptual organisation. Functional analysis consists of thoroughly understanding the procedures and processes that make the client's organisation work. This understanding is the basis for developing an approximate proximity matrix. Considerations of the number of people, the number of trips between spaces and the desired level of chance encounters between client groups can be factored into the adjacency matrix which in turn is developed into a graphic diagram of space adjacency.

The project briefing document is the record of all the decisions and agreements made about the inception of the project, scope, the data associated with the project, the mission, the intentions for quality, the functional requirements, and the preliminary conceptual proposals that give direction to the project. The document should be carefully crafted to speak to its intended audience whether this is the community, the financiers, or the project team and the client representatives. This implies a language that can be readily understood by the intended reader.

The stages of the project brief can be summarised as follows:

(1)   Review of the mission, the meta-goal and the subsidiary goals of the project.

(2)   It may well be the case that the strategic brief is prepared and the decision to build is taken by client members who are at a senior level in the hierarchy rather than by those subsequently responsible for the details contained in the project brief. The facts, issues, value system, mission and goals will therefore be rehearsed with the project briefing team.

(3)   Listing the users. The client's value system having been fully understood, the attention of the team is directed towards considering all of those user groups who will use the facility. The user list should not be confused with the stakeholder list although it will often be extensive and will require grouping – for example clerks, switchboard operators, secretaries, may all use the building in the same way and can therefore be grouped under a heading of administrative staff. In the example project, the users of the visitors centre are:

- head ranger
- assistant rangers
- administrative staff
- shop staff
- kitchen and snack bar staff
- cleaning staff
- maintenance staff
- visitors

3.18   Flowcharting and space identification

For each user group a flowchart is prepared which describes, in functional terms, the activities undertaken by each user. Each event on the flowchart represents a functional space requirement. Many of the flowcharts will incorporate common space where two or more functional activities can take place within the same space. An example flowchart is given in Fig. 3.4. It should be noted that each activity including movement between activities requires space, namely:

- Staff entrance
- Head ranger's office
- Briefing room
- Staff changing room and toilets
- Staff mess room
- Administrative office

Other space will be revealed by the flowcharts of other users' activity.

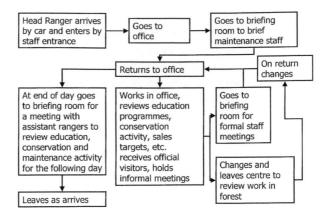

Fig. 3.4   Flowchart of head ranger's activities.

3.19   Adjacency

Having listed all of the spaces it is necessary to determine the relationship of the spaces one with another. This is done through the construction of an adjacency matrix (Fig. 3.5). The adjacency matrix looks similar to a conventional mileage chart but instead of the mileages shows relationships on a scale of +5 to −5. If two spaces are required to be adjacent to one another then a score of +5 will be entered at their intersection. Conversely, if spaces are required not to be adjacent to one another then a score of −5 will be entered at their intersection. If it does not matter whether spaces are adjacent or not a score of zero will be entered at the intersection. Adjacency in this context means that the spaces are accessible one from the other. In an extreme circumstance (+5) only a door or an opening is between one space and the next. Spaces are seen to be not adjacent if there is no clear route from one to the next. This is an important

Staff entrance

−5	Visitor entrance						
−5	+5	Visitor toilets					
+4	+3	−5	Ranger's office				
+3	+2	−5	+5	Administrative office			
+4	−2	−5	+4	0	Briefing room		
+5	−5	−3	−3	−2	+1	Staff change and toilets	
+2	−5	−5	−3	−2	+4	+1	Staff mess room

Fig 3.5   Adjacency matrix of selected spaces within the visitors centre.

concept, as spaces divided by a wall without a clear route from one to the other are considered not adjacent. Under this rule, spaces required not to be adjacent on a scale of -5 may be geometrically next to each other but may be insulated from one another in respect of sound, heat and light transmission.

3.20   Performance specification of space

The space is specified in terms of a functional statement of the physical size, quality, technology requirements, and environment. The size of the space will be stated in functional rather than measurement terms. The quality of the space will be defined generally in terms of its finishes and furnishings. The technology requirements of the space relate to the extent to which the spaces contain IT, communications, audio and video equipment, gases and other specialist services. The environment specification relates to the heating, ventilation, air treatment and acoustic requirements.

In the example project a description of the head ranger's office might be:

- Office of sufficient size for the head ranger to carry out administrative work with the use of a computer and hold formal and informal meetings with up to six other people.
- The office to be comfortable and of good quality but with care taken with regard to the environmental impact of the fittings, finishes and furnishings chosen.
- The office to have sufficient and protected power provision to support a personal computer with broadband internet connection, printer, scanner, paper shredder, voice mail system, telephone, video and up to three other items of powered office equipment. Sufficient power to be provided for free-standing lighting.
- The office will be naturally ventilated with temperature to be contained within the range 68°F to 74°F (20°C to 24°C). The office should be predominantly naturally lit with Cat. 2 lighting above the working area and formal meeting area; lighting to be muted in the informal meeting area. Encouragement is given to innovative but reliable solutions to the maintenance of a comfortable working environment with the minimum of energy usage. The acoustic characteristic of the office should be one of quiet with forest sound rather than human sound.

3.21 Characteristic of the final document

The project brief will contain the project mission and goal descriptions from the strategic brief along with the data derived above. The brief should impart to the design team all of the information necessary for the designers to understand the position of the project in the corporate strategy of the client as well as the basic performance criteria required by the project. If all of the above data are given and fully understood by the design team then the final design will answer the client's needs and desires in appropriate balance.

References

Banwell Report (1964) – *see* Ministry of Public Building and Works.

Barrett, P. & Stanley, C. (1999) *Better Construction Briefing*. Blackwell Science.

Construction Industry Board (1997) *Briefing the Team*. Thomas Telford.

Duerk, D.P. (1993) *Architectural Programming: Information Management for Design*. Wiley.

Kelly, J., MacPherson, S. & Male, S. (1992) *The Briefing Process: A Review and Critique*. RICS.

Latham, Sir Michael (1994) *Constructing the Team* (the Latham Report). HMSO.

Luft, J. & Ingham, H. (1955) *The Johari Window: A Graphic Model for Interpersonal Relations*. University of California Western Training Laboratory.

Morris, P.W.G. & Hough, G.H. (1987) *The Anatomy of Major Projects: a study of the reality of project management*. Wiley.

Ministry of Public Building and Works (1964) *The Placing and Management of Contracts for Building and Civil Engineering Work* (The Banwell Report). HMSO.

Pena, W.M., Parshall, S.A. & Kelly, K.A. (1987) *Problem Seeking: an architectural programming primer*, 3rd edn. AIA Press.

4 Benchmarking

David Eaton

4.1 Introduction

Benchmarking is the application of the skill of comparison – comparing one's own performance of a particular strategy, operation, task or operation with someone else's. In the construction industry the comparator should be someone who's past performance has been good. The term benchmarking has been adopted when the comparator is regarded as the best available. Benchmarking compares a snapshot of performance at a particular instant in time. This snapshot indicates the current performance and the inherent potential for future performance. The benchmarking activity should be stakeholder driven. Commitment to such fundamental review of processes and practices requires acceptance by the stakeholders. The activity has two uses: firstly, to generate quantitative measures of performance; and secondly, as a method of highlighting qualitative data relating to performance.

The typical response to a benchmarking exercise is the alteration of structure or strategy or systems or behaviour. There are four alternative strategic objectives to consider:

- Paradigm shifts
- Incremental change
- SWOT identification (strengths, weaknesses, opportunities, threats)
- CSF identification (critical success factors)

The construction industry can learn from benchmarking by having a greater awareness of targets and how best practice organisations achieve their goals. It can benefit from others' experiences by using them to innovate.

4.2 Benchmarking – helping yourself to be better

The principal purpose of benchmarking is to allow an organisation to compare its performance over a range of criteria with other organisations. This comparison may be necessary to: fulfil the organisational mission statement; allow the organisation to grow or to develop to be the best that it can be. It typically relates to meeting organisational aspirations. Benchmarking is the application of the skill of comparison – comparing one's own performance of a particular strategy, operation, task or operation with someone else's. In the construction industry the comparator should be an organisation whose past performance has been good. The term benchmarking has been adopted when the comparator is regarded as the best available. Benchmarking has no new techniques, no toolkits, pro-forma or software. It is quite simply a view of:

'... how well the future business compares with its competitors, not only in its likely market share, but in things like the rate of introduction of new products and services, and by counting the development of intellectual assets as an investment.'

Handy (1995)

Thus benchmarking is concerned with the comparison of the detail of the organisation. How its systems, structure, strategy and behaviours are manifested within the organisation, how they are promoted by the distinctive capabilities of the organisation in delivering services to clients and how they monitor and respond to external environmental influences. It is also concerned with the dynamic environment. How the organisation can utilise its systems, structure, strategy and behaviours to respond to changes in the competitive marketplace and hence change over time.

Benchmarking compares a snapshot of performance at a particular instant in time. This snapshot indicates the current performance and the inherent potential for future performance.

'At each [snapshot] there is found a material result: a sum of productive forces, an historically created relation of individuals to nature and to one another, which is handed down to each [snapshot] from its predecessor; a mass of productive forces, capital funds and [factor] conditions, which on the one hand, is indeed modified by the new [snapshot], but also on the other prescribes for it its conditions of life and gives it a definite development, a special character.'

Karl Marx (1972)

Benchmarking does not itself change practices or procedures; other measures are necessary to implement changes, but it does create the momentum to kick-start the change.

4.3 The origins

The origins of benchmarking within the construction industry arose from the construction 'quality' improvement movement, in the form of quality assurance and total quality initiatives. However, the origins of benchmarking lie beyond the construction industry. Comparisons have long existed. The treatment of the prodigal son raised biblical questions of comparison. The Taipan of the Princely House Hong becoming the most significant trader in Hong Kong resulted in the Princely House being used as the benchmark for business and finance comparisons throughout the Far East.

Yet the mathematical quantification of such comparisons has a much more recent commercial origin. In the late 1970s the American Xerox Corporation had to respond to a dramatic loss of market share to Japanese competitors and in particular the Canon Corporation (Camp, 1989). Xerox found that the *kaizen* (incremental improvements) developments had rapidly led to a significant alteration in the 'dynamic' and 'detail' environments (Eaton, 2000) of the competitive market for its copier equipment. This meant that there had been a significant regression in the sources of competitive advantage held by Xerox. Crudely put, Xerox equipment was technically inferior to its Japanese competition, more highly priced than the competition, and sold with a lesser technical support and repair facility. Canon could sell a comparable copier for less than Xerox's manufacturing costs.

The response from Xerox was to embark on a radical restructuring of the entire organisation to counter the threat to its existence. The question then arose as to what form the new structure should take. Xerox's conclusion was that the organisational structure should emerge together with those of the strategy, systems and behaviours of the corporation as they responded to the Japanese threat. Each element of structure, strategy, systems and behaviour would be established by comparison with the 'best available' comparators. In this particular instance Xerox sought to achieve a comparison with the 'best available competitor', namely the Canon Corporation.

Xerox established performance benchmarks for Canon, and then set about altering each element of structure, strategy, systems and behaviour to match or better the Canon target. Within a few years Xerox had reduced its cost per item sold by 50%, had brought its quality control and defects to levels comparable with Canon, and had radically altered the after-sales service and repair behaviour within the organisation. Xerox had survived the competitive crisis. The business had been transformed by competing against benchmarked targets. A cautionary warning is also provided by the Xerox experience. Kearns and Nadler (1992) state: 'One mistake we did

make was the idea of having three objectives equal in importance. . .
To affect a culture change. . . you have to zero in [on one issue only].'

Since then, many organisations have used benchmarking comparisons to stimulate responses to changes in quality improvement initiatives (e.g. British Airports Authority: Bacon, 1997; Construct IT, 1997), cost reduction strategies (e.g. Rover Motor Group: Graves & Madigan, 1997; Construct IT, 1997), budget allocation systems (e.g. BAe), new business-venture evaluation (e.g. British Telecom), ethical and cultural behavioural changes (e.g. The Body Shop), environmental and sustainability system alterations (e.g. Degussa AG and Skanska AB).

4.4   Benchmarking: a definition

Benchmarking can provide an objective analysis of how successful is an organisation's performance. It aims to adopt a systematic measurement process of the improvement to an organisation's performance in the utilisation of inputs, transformation system and outputs (in the form of its product or service) and to compare these measurements against the following:

- The organisation's own vision of performance
- The organisation's main rivals' level of performance
- The 'best available' comparators from other industries

The measurement of performance will result in quantitative and qualitative data that indicate the gap in performance. Management must then initiate change to alter performance. Such alterations, as previously indicated, may be changes in structure, strategy, systems or behaviour. The benchmark objectives can therefore vary from:

- Internal detail comparisons
- Competitors' internal detail comparisons
- Competitors' distinctive capabilities comparisons
- Competitors' external detail comparisons
- Industry-wide external detail comparisons
- Out-of-industry external detail comparisons
- Internal dynamic environment comparisons
- Competitor's dynamic environment comparisons
- Industry-wide dynamic environment comparisons
- Out-of-industry dynamic environment comparisons

Careful consideration is required to determine the appropriate objective and therefore the appropriate methodological approach to benchmarking to achieve prescribed objectives. The objective of construction benchmarking is the appropriate application of a tool (benchmarking techniques) to satisfy a management vision of construction business process improvement, thus aiming to implement continuous improvement (*kaizen*) of all business activity to add value and to counter the natural regression of sources of competitive advantage.

4.5   Prerequisites of benchmarking

The benchmarking activity should be stakeholder driven. Commitment to such fundamental review of processes and practices requires acceptance by the stakeholders. Reservations, restrictions and exclusions should be avoided. Nothing should be exempt from analysis, since this may lead to fragmentation and false assumptions and ultimately to a failure to achieve an accurate measure of the issue.

Peter Senge (1990) noted in his book, *The Fifth Discipline*, 'that people do not resist change, they resist being changed.' Thus without the participation of all, the benchmarking procedure may not yield accurate data, and therefore the benchmarking targets may be fundamentally flawed. The procedure must be forward-looking and focused on internal visions. Benchmarking allows the organisation to move from where it is to where it wishes to be. It can also be used to identify what may be necessary to move the organisation forward. However, benchmarking cannot tell the organisation where it should want to be nor should it be used to identify how an organisation arrived at its current position or identify those 'responsible'.

Therefore the prerequisites for benchmarking are that the aims and objectives of the benchmarking procedure should be made explicit to the stakeholders and the entire procedure should be transparent. This involves a significant degree of trust and mutual respect between the stakeholders. It is also recognised that in many benchmarking exercises these prerequisites do not exist. Caution is recommended in the interpretation of any results of a benchmarking exercise that fails to satisfy the criteria of necessary prerequisites.

4.6   Benchmarking processes

The strategic objective for the benchmarking process should be predetermined. The impetus for a benchmarking activity will typically have been derived from an exogenous source. There are four alternative strategic objectives to consider:

- Paradigm shifts
- Incremental change
- SWOT identification
- CSF identification

For example, a gradual loss of competitive advantage within a market may require an incremental benchmark procedure. Alternatively, the elimination of a market, due to a change in government policy, may require a paradigm shift benchmark procedure. The SWOT identification could be required as part of a strategic review of current performance. The CSF identification could be a part of the tactical implementation of a particular strategy. Whatever the exogenous prompt for change is, the correct process must be selected to enable the strategic objective to be achieved. When the strategic objective has been defined then the type of benchmarking process can be selected (Fig. 4.1). The alternative benchmarking types are (Camp's (1989) criteria amended and supplemented):

	Paradigm	Incremental	SWOT	CSF		
Internal	+/−	*	*/+	*	*	Typical
Competitive	*/+	*	*	*	+	Possible
Parallel	+	+	×	×	−	Unusual
Best practice	*	*	*/+	*	×	Impractical

Fig 4.1   Appropriate benchmarking methodologies.

- Internal: Best-in-class groups from within the organisation.
- Competitive: current direct competitors.
- Parallel: latent competitors, including those in the same industry but not in the same market, or those from other industries with a substitute product.
- Best practice: best in class from same industry or other industries.

These types of benchmarking can be undertaken with or without the cooperation, consent or knowledge of the parties involved. In the context of benchmarking other organisations there appears to be four alternative approaches:

(1) Competitive – hostile: where the organisation under examination does not assist the exercise and may try to subvert the results.

(2) Competitive – consensual: where the organisation does not assist the exercise but accepts that comparison is inevitable.

(3) Co-operative: where knowledge is exchanged, although the veracity of the information is unknown.

(4) Collaborative: where there is full knowledge exchange and verification mechanisms exist.

When considering the use of a particular benchmarking process it is important to recognise the constraints inherent in the procedures. Typical constraints would concern the anonymity of respondents when the process involved questionnaire, survey, case study or action research. Internal benchmarking assumes a level of cross-functional knowledge and experience that may not exist. When using a parallel or best practice procedure the selection of the comparator will depend upon the eclectic criteria of selection: how exhaustive should the search be for a 'best' comparator?; how exhaustive should the available data be for the use of the selected comparator? Figure 4.2 refers.

	Questionnaire	Interviews	Case studies	Action research		
Internal	*	*	*/+	–	*	Typical
Competitive	+	×	×	–	+	Possible
Parallel	*/+	–	–	×	–	Unusual
Best practice	+	+	×	×	×	Impractical

Fig 4.2   Appropriate benchmarking techniques.

4.7   Appropriate benchmarking techniques

Responses to appropriate benchmarking processes can include quantitative and qualitative data, the treatment of which requires the use of different techniques. A trend in current research has been towards reductionism, for example, Newtonian scientists break things apart and look at them one at a time. In physics, a fundamental presumption on the way to understanding the world is to keep isolating its ingredients until fundamental parts can be understood. There is a presumption that the other parts, which are not understood, are mere details. Benchmarking tries to capture both the dynamic and detail complexity of real world problems and therefore a different philosophical approach of looking at the whole is needed as a necessary precursor to the effective use of benchmarking. The Newtonian reductionism approach appears in many benchmarking exercises as the more holistic approaches have a more complicated research methodology and less clearly defined and interpretable results. For an appreciation of the reservations and constraints associated with the use of the various benchmarking approaches reference should be made to research methodology texts.

4.8   Benchmarking methodology

Whatever process is selected, the methodology follows a consistent pattern as listed below (Boxwell's (1994) basic steps are identified in italics):

- Identification of the core issues under scrutiny: *deciding what to benchmark and planning how to benchmark.*
- Internal data collection: *understanding organisations' own performance.*
- External data collection: *studying other organisations.*
- Analysis of data: *learning from the data.*
- Production of conclusions.
- Implementation of responses to the conclusions: *using the findings.*
- Feedback loop.

This methodology requires a specific skills mix and therefore careful selection of the person to conduct the benchmarking exercise is necessary. The process will involve the analysis and interpretation of quantitative and qualitative measurement. The resultant conclusions can only be as good as the skills of the investigator.

Charles Handy (1995) identifies nine different 'intelligences' (or skills), and a review of an effective construction benchmark researcher using the appropriate methodology and procedure would require all except musical skills. The nine 'intelligences' are:

(1)   Factual
(2)   Analytical
(3)   Linguistic
(4)   Spatial
(5)   Musical

(6) Practical
(7) Physical
(8) Intuitive
(9) Interpersonal

Apart from musical 'intelligence', the others would be used in conceptualising, coordinating, consolidating and communicating the benchmarking procedure.

Camp (1989) identifies ten success metrics that ensure a benchmarking exercise runs effectively:

(1) An active commitment to benchmarking from senior management.
(2) A clear and comprehensive understanding of how one's own work is conducted as a basis for comparison to industry best practices.
(3) A willingness to change and adapt based on benchmark findings.
(4) A realisation that competition is constantly changing and there is a need to 'shoot ahead of the duck'.
(5) A willingness to share information with benchmarking collaborators.
(6) A focus on benchmarking, first on industry best practices and secondly on performance metrics.
(7) The concentration on leading companies in the industry or other functionally best operations that are recognised leaders.
(8) Adherence to the benchmarking process.
(9) Openness to new ideas, being creative and innovative in the application of new procedures to existing processes.
(10) The institutionalisation of benchmarking.

Watson (1993) considers the ethics of benchmarking, and draws the following conclusions:

● All information-gathering should be within legal limits.
● All information should be fully shared between benchmarking collaborators and should be kept confidential if requested.
● The expectations of the collaborating parties are fully understood, both within the research team and when in contact with external parties.

4.9 The uses of benchmarking: performance measurement

The benchmarking process has two uses: as a method of highlighting qualitative data relating to performance and to generate quantitative measures of performance which may be utilised to create metrics and milestones for comparing against exemplars of 'best practice'. The qualitative data are more usually used in support of the development of strategies for some form of quality improvement within the organisation, for example operational improvements associated with the introduction of life-cycle costing or facilities management procedures. The qualitative data from competitor organisations can be used to promote cost reduction or budgetary control improvements. Such data can also be used in the evaluation of new ventures or as a confirmation of proposed action in response to a competitive crisis. Thus the effective use of quantitative and qualitative data can typically inform corporate visions and corporate plans and substantiate and justify proposed strategies. These data can also be used to confirm the tactical necessity for the implementation of revised procedures and controls.

Alternatively, a more efficient use of benchmarking is to facilitate organisational learning and organisational change. This form of use is a complex issue taking the organisational response beyond what is typically a reactionary response and into generative learning, including the implementation of both single- and double-loop learning in which the organisation can identify the structural explanations of the underlying causes of behaviour (Segil, 1996). Altering the underlying structures can produce different patterns of behaviour in members of the organisation.

The potential application of qualitative benchmarking procedures exists in the evaluation of new and innovative developments within the construction industry. For example, recent developments in the creation and development of 'relational contracting' and 'partnering agreements' have created new contractual and 'quasi-contractual' relationships, the performance of which has yet to be evaluated. Such relationships include:

- PPP (public and private partnership initiatives)
- PFI (private finance initiative) schemes
- JVs and SPVs (joint ventures and special purpose vehicles)
- Partnerships
- Concords
- Value chain partnering agreements
- Lean construction *keiretsu* arrangements
- Federations
- Subsidiarity agreements

Similarly, human resource changes involving team working and empowerment initiatives will require an evaluation mechanism.

Benchmarking techniques can be utilised in this evaluation, for example in external processes:

- *Client management*: benchmarking the effectiveness of management systems by measuring, for example, the rate of client attrition, rate of new client introductions, proportion of turnover associated with new clients.
- *Consultant management*: benchmarking the efficiency of management systems and data manipulation by measuring, for example, ratio of CAD/CAM projects to traditional design, proportion of contract work outsourced, ratio of introduction of new consultants, average age of appointments to consultants' approved list.
- *Contractor management*: benchmarking the efficiency of management systems and data manipulation by measuring, for example, proportion of EDI (electronic data interchange) tendering, ratio of introduction of new contractors, ratio of recommendation to tender list, ratio of successful to unsuccessful bids.
- *Site process alteration*: benchmarking the effectiveness of management systems by measuring, for example, drawing revisions issued, method statement alterations, issue of variation orders, issue of verbal instructions, requests for clarification of details, receipt of notices for prolongation and disruption.
- *Briefing and design*: benchmarking the effectiveness of management systems by measuring, for example, client feedback, statutory authority compliances, environmental impact assessments, BREEAM assessments, design reviews, proportion of projects using 3D modelling, use of virtual reality walk-throughs.
- *Planning, monitoring and control*: benchmarking the effectiveness and efficiency of management systems and data manipulation by measuring, for example, resource levelling, progress report status, safety policy amendments, alteration to QA plans and reports.
- *Facilities management*: benchmarking the effectiveness and efficiency of management systems and data manipulation by measuring, for example, property acquisition ratios, changes of use, building usage, occupancy rates, lease management, space utilisation, budgetary control failures, risk management failures, life-cycle cost alterations.
- *Cost estimating and bidding*: benchmarking the effectiveness of management systems and data manipulation for budgeting, risk assessment, value management, cost modelling, site surveying, measurement, pricing and project planning.
- *Cost and variation management*: benchmarking the effectiveness of management systems and data manipulation for measurement,

pricing, cost planning, cost checking, life-cycle costing, value management.

- *Contract management*: benchmarking the effectiveness of contract management and data manipulation by measuring, for example, extensions of time for prolongation or disruption, payments, debtors, financial claims for loss and expense.
- *Tendering, enquiries, quotes and orders*: benchmarking the effectiveness of management systems and data manipulation for preparing, evaluating and completing enquiries, quotes and orders.

For internal organisational processes, benchmarking could be used for evaluation and amendment of:

- *Strategy*: benchmarking the effectiveness, efficiency and opportunity for, alteration of business and operational aims and objectives, quantification of particular business and operational contributions, extent of intra-organisational information transfer, extent of inter-organisational information transfer, ratio of use of IT to traditional communication, etc., new opportunities and business ventures.
- *Policy*: benchmarking the effectiveness, efficiency and opportunity of the current strategy, mission statement and vision.
- *Procedures*: benchmarking the effectiveness, efficiency and opportunity for, process improvement, application of new philosophy tools and techniques, staff development and training.
- *Communication*: benchmarking the effectiveness, efficiency and opportunity for innovation, introduction of EDI, intranet, e-mail, e-commerce, web data, etc.
- *Project-specific requirements*: benchmarking the effectiveness, efficiency and opportunity for creating niche and focused services utilising generic concepts and initiatives, spin-off services, spin-off facilities.
- *Selection of suppliers*: benchmarking the effectiveness, efficiency and opportunity for process improvements, partnering agreements, value chain improvements, *keiretsu* agreements, etc.
- *Legal and health and safety requirements*: benchmarking the effectiveness, efficiency and opportunity for performance improvements, covering safety policy, COSHH/CDM regulations, insurance, building regulations, British Standards and Codes of Practice.

4.10   Responses to the benchmarking process

The typical response to a benchmarking exercise is the alteration of structure, strategy, systems or behaviour. Benchmarking cannot achieve such changes; the process merely indicates what changes may be necessary. A more holistic and comprehensive alteration is when all four elements change simultaneously. Xerox is an exemplar of best practice organisational change. Typical examples of responses to benchmarking are indicated below.

(1) Structural change
 • Lean organisations
 • Empowered organisations
(2) Strategic change
 • Product differentiation: specialisation and additional services
 • Process differentiation: PFI procurement
 • Market differentiation: specialisation and niche focus
(3) Systemic change
 • Backward value chain integration: design and construct
 • Forward value chain integration: facilities management
(4) Behavioural change
 • VFM (value-for-money) initiatives
 • Value-added initiatives
 • Process re-engineering

4.11   Benefits of benchmarking

The changes implemented as a result of the benchmarking process yield benefits associated with organisational effectiveness, efficiency and opportunity:

(1) Competitive advantage via:
 • Integration
 • Flexibility
 • Speed: improved cycle times
 • Quality: reduction in defects
 • Cost reductions
 • Customer focus: delivering client expectations
 • Improved image and reputation
(2) Profitability via:
 • Increased opportunity
 • Cost reductions
 • Reduced wastage
 • Reduced re-work

(3) New markets via:
- Differentiation
- Focus
- Niche

(4) New products and services via:
- Segmentation
- Specialisation

4.12   Implications of benchmarking

Many conclusions can be drawn from the results of a benchmarking exercise. The organisation must recognise the necessity of adopting a holistic approach to organisational change indicated by the benchmarking exercise and accept the appropriate level of investment to effect such change. This investment may not necessarily be financial investment, but inevitably there is a need for significant investment in people.

Benchmarking is a tool for continuous improvement therefore embarking on a benchmarking exercise indicates the necessity for continuous benchmarking activity. As this has inevitable implications for staff and other resources, the successful continuity of benchmarking depends upon the process being made transparent and the results of the benchmarking activity should be made available to those affected, therefore reinforcing the necessity for the proposed actions.

4.13   Benchmarking case study

An M4I publication of May 1999, (http://www.m4i.org.uk) identified industry 'average' performance metrics for ten key indicators of construction industry performance. Whilst not being 'best' practice benchmarks, they provide some useful information for internal benchmarking and process improvement objectives.

The KPI (key performance indicators), their measures and current 'averages' are:

- Construction cost – inflation adjusted improvement for the year-on-year cost of a given project: -3%.
- Construction time – year-on-year comparison of improvement in delivery time: -1%.
- Predictability – design cost – difference between design cost at approval stage and completion: 0.0%.

- Predictability – design time – estimated design time compared with actual time: 18%.
- Defects – scale of 1 to 10 (1 = totally dissatisfied): 8.
- Client satisfaction – product – scale of 1 to 10 (1 = totally dissatisfied): 8.
- Client satisfaction – service – scale of 1 to 10 (1 = totally dissatisfied): 8.
- Profitability – pre-tax profit as percentage of sales: 3.2%.
- Productivity – turnover per full-time equivalent employee: £60 000.
- Safety – reportable accidents per 100 000 employees per annum: 997.

M4I is suggesting that KPIs should become an essential pre-qualification measure for future projects and only those companies that have consistent KPI improvement would pass screening for the bid phase of projects. Some clients have already established a contractual requirement for KPIs, and others, such as ASDA, Railtrack and Severn & Trent Water Authority have minimum acceptable thresholds for each KPI. Benchmarking of 'hard' targets is becoming commonplace, with construction organisations benchmarking the activities indicated in this report.

'Egan' metrics have been published and represent:

- 10% annual savings in construction costs
- 10% annual savings in construction duration
- 20% annual reduction in project defects

Future developments of benchmarking applications will move to the measurement of 'softer' behavioural activities, and assessment of future business potential, for example:

- *Education and training needs:* evaluating the growing expectations of customers and staff, accounting for business growth through personal development and accounting for the potential loss if customers and/or staff become dissatisfied.
- *Professional scope and values:* achieving business development by promoting entrepreneurial initiative and by the adherence to professional standards.
- *Professional ethics:* assessing the impact of inappropriate professional behaviour and the likely reaction of professional bodies.
- *Professional standards:* evaluating the impact of failure to deliver services of an appropriate professional standard, to reduce the impact of potential claims for professional negligence.

- *Leadership effectiveness:* evaluating the effect of releasing the organisation's potential through effective management.
- *Customer focus*: to exceed client expectations and to minimise the impact of failure to meet client expectations which could cause potential damage to future business prospects or damage to image and reputation.
- *Integration of processes*: to support improvements in the whole construction life cycle through design, construction and facilities management, with the aims of enhancing value for the industry's clients and improving profitability for the industry.
- *Integration of teams*: to support the development of effective multi-disciplinary teams and to accrue the benefits from achieving a degree of permanency from such project teams.
- *Commitment to people*: for example by achieving recognition of employee development through schemes such as IIP (Investors in People) and CharterMark.

Some work is under way in achieving a balance in process improvement. A 'balanced scorecard' approach using a 'radar chart' representation is a possibility. Figure 4.3 indicates a fictitious 'radar chart', indicating performance as compared to the M4I KPIs.

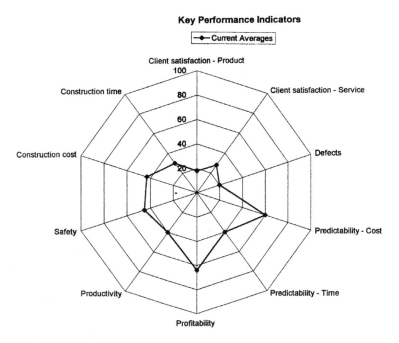

Fig. 4.3   M4I key performance indicators.

4.14   Other current research initiatives

There are a number of existing benchmarking initiatives that can be referred to for exemplars of good practice:

- Building Research Establishment's benchmarking network.
- Bath University's Agile Construction Initiative – benchmarking civil engineering projects.
- Salford University's Construct IT Centre of Excellence – benchmarking of IT implementation in construction.
- Human Systems Ltd – project management benchmarking network.
- Morrison Construction – balanced scorecard initiative.
- Movement for Innovation – M4I key performance indicators.
- European Construction Institute – CIB benchmarking projects.

4.15   Conclusions

The Egan report cites a construction industry representative's view after a visit to Nissan UK as:

> 'We see that construction has two choices: ignore all this in the belief that construction is so unique that there are no lessons to be learned; or seek improvement through re-engineering construction, learning as much as possible from those who have done it elsewhere.'
>
> (Egan, 1998)

The essence of benchmarking, learning as much as possible from those who have done it better elsewhere, will not alone improve anything within an organisation, but it does generate data that can be utilised in formulating strategy, policy and tactics. Holistic alteration of structure, strategy, systems and behaviour is undoubtedly necessary to improve performance within the construction industry. Benchmarking is a tool that can be utilised by management to foster such improvement. It will however require significant changes in the trust and respect shown to stakeholders within the construction industry.

The message is clear, the construction industry can learn from benchmarking. The construction industry can benefit from others' experiences; it can use these experiences to innovate. Having conducted a benchmarking exercise, the organisation will have a greater awareness of targets and of how 'best' practice organisations achieve their goals. Sufficient knowledge should then exist to make a difference in practice generally.

References

Bacon, M. (1997) 'Lean Construction: Process Change in BAA', *The Armathwaite Initiative – Global Construction IT Futures*, international meeting held 16–18 April 1997 at Armathwaite Hall, Bassenthwite, UK.

Boxwell, R.J. (1994) *Benchmarking for Competitive Advantage*. McGraw-Hill.

Camp, R.C. (1989) *Benchmarking: The Search for Industry Best Practices that Lead to Superior Performance. Quality Press*.

Construct IT (1997) *The Armathwaite Initiative report*. Construct IT Centre of Excellence, University of Salford.

Eaton, D. (2000) *A 'Detail' and 'Dynamic' Competitive Advantage Hierarchy Within the Construction Industry*. CIB W92.

Graves, A. & Madigan, D. (1997) 'The Experience of Benchmarking in Automotive and Aerospace Industries', *The Armathwaite Initiative – Global Construction IT Futures*, international meeting held 16–18 April 1997 at Armathwaite Hall, Bassenthwite, UK.

Handy, C. (1995) *Beyond Certainty*. Hutchinson.

Kearns, D.T. & Nadler, D.A. (1992) *Profits in the Dark: How Xerox Reinvented Itself and Beat Back the Japanese*. Harper Collins.

Marx, K. (1972) *The German Ideology* (ed. C.J. Arthur). International Publishers Co.

Segil, L. (1996) *Intelligent Business Alliances*. Century Business Books.

Senge, P. (1990) *The Fifth Discipline: The Art and Practice of the Learning Organisation*. Century Business Books.

Watson, G.H. (1993) *Strategic Benchmarking: How to Rate your Company's Performance Against the World's Best*. Wiley.

5 Value Management

John Kelly and Steven Male

5.1 Introduction

Value management is being increasingly used in the UK, manufacturing, service and construction sectors to maximise the benefits of a project, to the project's sponsor or client, where the definition of project is the 'investment of resource for return'. Investment may be in terms of capital or other resource input and the return may be measured in social, economic or commercial terms. Value management has been defined as a proactive, creative, problem-solving service. It involves the use of a structured, facilitated, multi-disciplinary team approach to make explicit the client's value system using function analysis to expose the relationship between time, cost and quality. Strategic and tactical decisions are audited against the client's value system at targeted stages through the development and life of a project (Kelly and Male, 1993).

The exploration of the client's value system will allow the appraisal of decisions at a number of key levels, which can be related to the manufacturing, service and construction sectors:

- Strategic planning and business definition
- Project planning and the establishment of systems
- Service definition of component parts of the project
- Operations and use

5.2 The development of value management

Value engineering has its foundation in the manufacturing sector of North America (Kelly and Male, 1993; Shillito and De Marle, 1992; Norton and McElligott, 1995). Initially called value analysis, the term value engineering is the commonly used term in North America and was formalised in the title of the Society of American Value Engineers, which recently changed its title to SAVE International. The value engineering concept began in the late 1940s when shortages of strategic materials forced, initially GEC, to consider alternatives that performed the same function. It was soon discovered that many of the alternatives provided equal or better quality at a reduced cost, leading to the first definition of value analysis, being:

> Value analysis is an organised approach to providing the necessary functions at the lowest cost.

Value analysis was always seen to be a cost validation exercise that did not affect the quality of the product. However, it was recognised that many products had unnecessary cost incorporated by design. This led to a second definition of value analysis, being:

> Value analysis is an organised approach to the identification and elimination of unnecessary cost where unnecessary cost is defined as a cost which provides neither use, nor life, nor quality, nor appearance, nor customer features.

From the beginning, value analysis was seen as a team exercise.

In 1954 the US Department of Defense bureau ships became the first US government organisation to implement a formal programme of value analysis. It was at this time that the name change to value engineering was made for the administrative reason that engineers were considered the most appropriate personnel to undertake the task. This led to an extensive development of value engineering through the US public sector, culminating in section 36 of public law 104–106, which defines value engineering and requires its use within the public sector.

Value engineering in the UK began in the 1960s manufacturing sector and led to the establishment in 1966 of the Value Engineering Association. This organisation changed its name in 1972 to the Institute of Value Management. The term value management is in common use throughout Europe (except France were the term value analysis is used) to describe this service. Value engineering is commonly seen to be a subset of value management as shown in Fig. 5.1. Value engineeering opportunities commence at the stage when the project can be defined in terms of its elements and components and terminates upon the completion of the physical project. Value management opportunities commence at the inception of the project and continue through the project's life.

Value management applied to construction became popular in the UK in the early to middle 1990s and has led to a number of guidance documents (BRE, 1997; CIB, 1997; CIRIA, 1995; European Commission, 1995; HM Treasury; ICE, 1996) British Standard BS EN 12973: 2000 Value Management was published in June 2000. Although

CLIENT'S BUSINESS/CORPORATE STRATEGY IMPACTING THE PROJECT

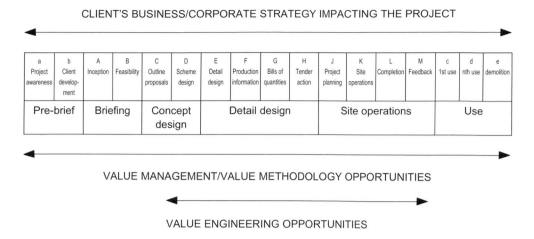

a Project awareness	b Client develop- ment	A Inception	B Feasibility	C Outline proposals	D Scheme design	E Detail design	F Production information	G Bills of quantities	H Tender action	J Project planning	K Site operations	L Completion	M Feedback	c 1st use	d nth use	e demolition
Pre-brief		Briefing		Concept design		Detail design				Site operations				Use		

Fig. 5.1   **Opportunities for value management and value engineering.**

VALUE MANAGEMENT/VALUE METHODOLOGY OPPORTUNITIES

VALUE ENGINEERING OPPORTUNITIES

adopted enthusiastically by the UK private sector, the public sector has demonstrated less interest. However, the requirements of best value in the local authority sector and the introduction of prime contracting as a key procurement route has led to increased interest by the public sector. Prime contracting requires that value management be used at all stages of the development of the project and involves a team comprising the client, users of the proposed facility, the design team, the prime contractor and cluster leaders responsibile for significant parts of the technical development of the project.

The increasing use of value management has led to a more formalised view of the service in terms of the commonly used tools and techniques. It is these that will be described in outline in this chapter.

5.3   An outline of value management

A generic structure for a value management exercise is represented in Fig. 5.2. There are various triggers for a value management exercise; examples might be new legislation, a new opportunity for a commercial product, the solution of a social problem, the planning of a complex construction operation or simply an overspend on budget needing to be addressed. The trigger indicates project awareness and starts the process of client development.

It is presumed here that value management exercises will be workshop-based and that the workshop team will have all necessary information for the recognition of the problem to be determined and addressed. This information may be through the knowledge of the

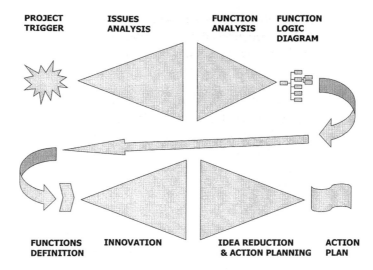

PROJECT   **ISSUES**   **FUNCTION**   **FUNCTION**
TRIGGER   **ANALYSIS**   **ANALYSIS**   **LOGIC**
                                           DIAGRAM

FUNCTIONS   **INNOVATION**   **IDEA REDUCTION**   **ACTION**
DEFINITION   **& ACTION PLANNING**   **PLAN**

Fig. 5.2   A generic model of the VM process.

team members present or by hard data brought to the workshop. Not all value management is workshop-based; the French, for example, tend towards a system of value analysis which does not depend upon the workshop.

Whatever the trigger and whatever the stage in the project, the first stage in a value management exercise is for all members of the team to undertake an information-sharing and interpretation exercise and thereby generate the issues. The issues analysis is expansive in nature, as illustrated in Fig. 5.2, and is closed down through a two-part exercise. The first part considers which of the issues are of real importance to the project; the second part considers which are the 'show-stoppers', i.e. those issues which if not resolved will stop the project dead. All of those issues highlighted in this process are explored in detail and their application to the function of the project defined. This leads to the second stage of the value management exercise – function analysis.

Function is defined here as a characteristic activity or action for which a thing exists. Therefore something can be termed functional when it is planned primarily in accordance with the requirements of use rather than primarily in accordance with fashion, taste or even rules and regulations. Functions generally require an answer to the question 'Why are we providing this service?' or 'What is this component designed to do?' Therefore function can be addressing the provision of educational and health services to the under-fives or examining in detail the requirements of a window. In either case a useful discipline on the identification of function is a definition in as few words as possible and ideally through the use of two words – an active verb and a descriptive noun. It is useful to link all functions through some form of function logic diagram that explains the hierarchy within the functional approach to the project. This technique is illustrated in the final section of this chapter as an example.

The definition of function is an important step as, once realised, functions provide the foundation and catalyst for innovative solutions. For example, it is difficult to be innovative on alternatives to an internal wall without identifying exactly why that particular internal wall exists. Internal walls generally separate space, secure spaces, separate environments, maintain privacy, support fixtures, transfer load, attenuate noise, etc. Innovative solutions will depend upon the function of the particular wall under consideration. If the function of the wall is simply to delineate spaces then a row of pot plants or a change in the floor finish may suffice. The innovation stage in a team environment is most likely to be carried out through brainstorming. The brainstorming of proposals is an expansive exercise in a similar way to the issues analysis (see Fig. 5.1) and is normally followed by a structured reduction in the number of ideas resulting in a list of those that are clearly winning ideas.

An analysis of the winning ideas will lead to an action plan. This will require named members of the team to champion the development of an idea, which in the UK is often undertaken outside the workshop.

5.4   The lever of quality

The lever of quality is a diagrammatic representation of the effort required at the four generic stages in the development of a project, to achieve a measured increase in the quality of the product or service in accordance with the requirements of the client's value system (Fig. 5.3). It is easier to obtain a measured increase in quality by exerting pressure at the strategic planning stage rather than the operations stage. This argues for a structured programme of value management throughout a project such that pressure is exerted at every value opportunity point.

5.5   The international benchmarking of value management

The following sections incorporate the results of a research project undertaken by the authors to establish a benchmark of value management practice (Male and Kelly, 1998). The project related to practice in construction and was funded by the Engineering and

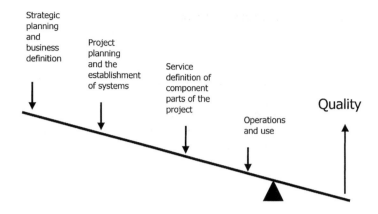

Fig. 5.3 The lever of quality (from an original idea and with acknowledgements to Winston Davis, Jaguar Cars Ltd).

Physical Sciences Research Council under the Innovative Manufacturing Initiative programme.

The project reviewed all available literature and involved international visits and questionnaire surveys. The benchmark described in this chapter is a factual exposition of the current practice of value management as it is applied, internationally, in the construction industry.

The research identified a series of critical success factors for VM which include:

- Multi-disciplinary team/appropriate skill mix
- The skill of the facilitator
- The structured approach through the 'VM process'
- A degree of VM knowledge on the part of the participants
- Presence of decision-takers in the workshop
- Participant ownership of the VM process output
- Preparation prior to the VM workshop
- Functional analysis
- Participant and senior management support for VM
- A plan for implementation

It was also noted from the research that all of the above success factors were quoted in the 'negative' as being the potential problems or pitfalls of the VM process. There was a high degree of concurrence on the use of untrained and inexperienced facilitators being identified as detrimental to both the outcome of the VM process and its future development as a value-enhancing tool for clients.

5.6   The VM process in detail

Opportunities for value management

In construction there was consensus on the four opportunities where the VM process could be employed to achieve maximum effect on any project during its life cycle. These are:

(1)   The pre-brief workshop
(2)   The brief workshop
(3)   The concept design workshop
(4)   The detail design workshop

It was acknowledged that stages (1), (2) and often (3) could be combined as a review of the brief in an exercise termed a 'charette'. Figure 5.4 illustrates the value opportunities with reference to the RIBA Plan of Work. In the research the four opportunities were found to correlate well with the generic points for action represented by the lever of quality.

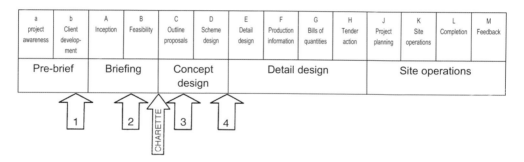

Fig. 5.4   VM opportunities with reference to the Royal Institute of British Architects modified Plan of Work.

Figure 5.5 gives a diagrammatic representation of the VM process.

Prerequisites for value management workshop success

The VM process prerequisites that must be satisfied to ensure a smooth-running workshop, relate to the involvement of people and the venue for the workshop. These include:

- Agreement by all parties to participate
- Senior management and participant support for value management
- Experienced facilitator(s)

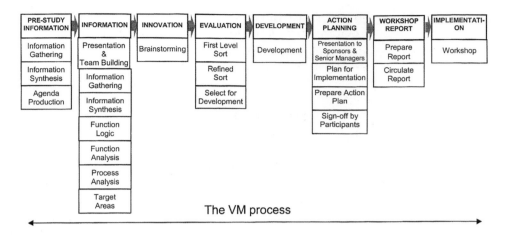

Fig. 5.5   The value management process.

- Appropriate team skill mix
- Presence of client decision-taker(s)
- Isolated workshop environment

Agreement to participate

Agreement to participate by all those attending the workshop is seen by many facilitators as fundamental to the success of the workshop. This may involve the facilitator interviewing individually or holding a pre-workshop meeting (orientation) with the aim of dispelling any nervousness of the process. Senior management sets the scene by endorsing VM to ensure that everyone knows that ideas and suggestions arising from subsequent workshops will be implemented. This signals to participants that what they say and do will have an impact on the project.

Facilitation

Experienced facilitators were cited by the majority of benchmarking partners as being paramount for a successful VM workshop. The facilitator is, in effect, a VM process manager and acts as team coordinator, guiding the team through the tools and techniques best matched to the project under investigation. Key skills were highlighted as being:

- Setting the workshop agenda.
- Managing the process of the workshop including:
  - Dealing with any hidden agendas of participants
  - Recognising individuals and their contribution

- ○ Questioning and summarising
- ○ Providing direction and a sense of common purpose
- ○ Sensing interpersonal relationships within the team
- ○ Sensing the climate of the workshop
- ○ Synthesising and integrating information during the workshop
- ○ Intervening in the workshop process when necessary
- Communicating verbally and producing the VM report.

In respect of the debate over external versus internal facilitator, the consensus was for an external facilitator, the definition of external being someone who is not explicitly involved with the project being analysed. External facilitators are considered to be more effective than internal due to the fact that they are likely to pose probing and challenging questions without fear of sounding non-knowledgeable. An external facilitator also has no vested interest or hidden agendas with regard to the project under workshop. The facilitator will:

- Provide direction and a sense of common purpose
- Sense interpersonal relationships within the team
- Sense the climate of the workshop
- Synthesise and integrate information during the workshop
- Intervene in the workshop process when necessary
- Communicate verbally and produce the VM report

Dual facilitation (two facilitators as opposed to one facilitator or one facilitator with a recorder) has emerged as the preferred option due to the mutual help the facilitators can give each other and also the variation it gives the participants. Two facilitators alternating can keep the momentum going to a higher degree than someone working on their own. Some stated that it is also possible to work longer days without sacrificing quality. However, while this was preferred, it was acknowledged that the economics of the marketplace tended to drive facilitators to quote for a single facilitator option or sometimes one facilitator with a recorder.

Team Issues

In the USA it is common to use a separate audit team to examine the project rather than use the project team. However, it was widely acknowledged internationally that the existing project team should be the basis of the VM workshop team, bringing in experts to the workshop if required. The advantages of using the existing team were quoted as being:

- Costs are kept down
- Better ideas may crop up due to greater experience 'second' time around
- There is another chance to explore alternative options
- VM workshops are useful for team building and resolving conflicts that may exist within the existing team
- Increased communication
- Increased implementation

Empowerment of team members

The empowerment of some of the participants of a VM workshop team to *take* decisions will influence the effectiveness of the workshop in terms of implementation of ideas. The term decision-*taker* is used here to represent a person who has the authority to make and take decisions during the workshop, whereas a decision-*maker* can only provisionally endorse solutions and needs to refer to a higher authority in order for the solutions to be ratified and then implemented. It is suggested that the most effective workshop, in terms of implementation, is one where the client representative has the authority to take decisions and is a member of the workshop team working with the project team. The least effective, in terms of implementation of ideas, is one where there is no client representative and the team is fully independent.

There are a number of issues that arise from this debate. Firstly, the role of the facilitator is influenced by the above relationships. With an existing team the facilitator has to manage the VM process and assume a challenging role in order to assist the team members to 'break' from their traditional thinking in the project context. On the other hand, for an independent team, the facilitator has to be more oriented towards managing the VM process only.

It should be noted that the presence of a client decision-taker may not be required for workshops conducted later in the project life cycle, as the client's value system should already be encapsulated in the brief and sketch designs. The value management process in this situation will be concerned with matters of performance of elements and components. However, it would be necessary to bring a client decision-taker on board if no VM workshops had been carried out on the project previously, and the workshop was likely to consider issues of project concept and spatial layout.

Workshop location

The likelihood of a VM workshop being successful is enhanced if an isolated environment is provided for the participants. An area should be set aside where the workshop can be held without interruptions from outside agents. Advantages cited include:

- It focuses the team on 'the project'
- Gestation occurs during the workshop process
- It commits the team
- Militates against partial attendance
- Continuity is ensured

It is important to weigh up the cost of providing the isolated environment to maximise the outcomes versus apparent cost. An isolated environment, for example in a hotel, does focus the team totally on the project under investigation in a 'pressure cooker' environment. Residential studies stimulate team building but workshop costs are higher than non-residential.

An analysis of the process highlighted a significant level of consensus amongst the international value management community. In benchmarking, very little time was spent explaining systems and almost none in explaining terms.

5.7   Workshop types

As stated above, in construction there is consensus on the four opportunities where the VM process could be employed to achieve maximum effect on any project during its life cycle. These are:

(1)   The pre-brief workshop
(2)   The brief workshop
(3)   The concept design workshop
(4)   The detail design workshop

Pre-brief workshop – the strategic brief

The strategic brief sets out the broad scope and purpose of the project and its key parameters including the overall budget and schedule. It should provide an output specification that explains in clear terms what is expected of the project. The project at this stage exists as 'the investment of resource for return' and does not necessarily imply construction.

The primary purpose of the strategic brief is to structure information in a clear unambiguous form to permit a 'decision to build' to be taken in light of all of the pertinent facts which include the output of brainstorming. The strategic brief should include:

- The project mission statement
- The project context
- Organisational structure
- Overall scope and purpose of the project
- Schedule, including phasing
- Statements on size and capacity requirements and functions to be accommodated
- A global capital expenditure budget and cash flow constraints
- Targets and constraints on operating expenditure and other whole-life costs
- Internal and external environmental requirements
- Technology to be incorporated or accommodated, including equipment, services, IT
- Quality requirements for design, materials, construction and long-term maintenance
- Expectations in response to the brief
- How the success of the project will be measured
- Statutory requirements

The workshop is likely to last between $\frac{1}{2}$ and 1 day and involves a facilitator and senior representatives of the client organisation only. Since it is important to hear the views of all stakeholders of the project, the numbers attending this style of workshop tend to be large – between 10 and 30 or even more. This requires innovative team management and may involve splitting into sub-groups for short periods to study assigned tasks.

It was stressed during the benchmarking interviews that the facilitator is required to be well-prepared for these studies, which may necessitate extensive pre-study interview and the preparation of 'information packs' for all participants. Indicative tasks at the workshop include:

- Presentation and team building
- Checklist or brainstorm issues, group and theme
- Stakeholder analysis
- Strategic time line
- Project driver analysis
- Time/cost/quality analysis
- Function analysis
- Function logic diagram
- Brainstorming alternatives
- Evaluation and development
- Presentations
- Plan for implementation
- Prepare action plan

- Sign-off by participants
- Facilitator to prepare and circulate report

Brief workshop – the project brief

The project brief is undertaken after the decision to build; it restates the strategic brief and converts it into a performance specification detailing:

- The aims of the design, including prioritised project objectives
- The site, including details of accessibility and planning
- The functions and activities of the client
- The structure of the client organisation
- The size and configuration of the facilities
- Options for environmental delivery and control
- Servicing options and specification implications, e.g. security, deliveries, access, workplace, etc.
- Outline specifications of general and specific areas
- A budget for all elements
- The procurement process
- Environmental policy, including energy
- The project execution plan
- Key targets for quality, time and cost, including milestones for decisions
- Method for assessing and managing risks and validating design proposals
- Room data sheets may also be prepared

The primary purpose of the project brief is to structure information in a form that can be readily understood by the design professionals and the contractor. The project brief should overtly represent the client value system.

The workshop lasts between one and two days and involves a facilitator, the client's project sponsor(s), who is likely to be a senior representative of the client organisation, together with the design team. The views of all stakeholders of the project have been recorded in the strategic brief and therefore the numbers attending this workshop tend to reduced – usually between 10 and 20.

Indicative pre-workshop tasks include:

- Interviews with client; the checklist
- Document analysis
- Site tour
- Questionnaire

- Post-occupancy evaluation (POE) reports of similar buildings
- Facilities walk through similar buildings

Workshop tasks include:

- Presentation and team building
- Review function logic diagram, stakeholder analysis, time/cost/ quality analysis and strategic time line
- REDReSS (reorganisation, expansion, disposal, refurbishment and maintenance, safety and security)
- SWOT (strengths, weaknesses, opportunities and threats) analysis of the project
- Checklist or brainstorm issues, group and theme
- Process flowcharting
- Functional space analysis resulting in identification of, for example, size, shape, quality, heat, light, acoustic attributes, IT requirements, future demands
- Spatial adjacency
- Brainstorming alternatives
- Evaluation and development
- Presentations
- Plan for implementation (the project brief)
- Prepare action plan
- Sign-off by participants
- Facilitator to prepare and circulate brief and action plan as a report

Brief review workshop – the charette

This is an alternative to the preparation of the brief through the value management process and involves the review of an existing brief commonly prepared by a client, project manager or architect. In the charette, all of the stages of the strategic brief and project brief workshops are combined into a single workshop lasting between one and three days. This style of workshop is common in Australia and the UK. One UK value manager stated that the vast majority of all workshops facilitated were of the charette type. One Canadian client stated that a charette was the best way of briefing the design team and ensuring that the team fully understood the client requirements.

Concept design workshop – outline sketch design (OSD)

The concept design workshop is a value review of the initial plans, elevations, sections, specification and cost plan of the proposed facility using the signed-off concept brief as a reference point. It is important to note that it is impractical to develop designs within a workshop environment; designs have to be prepared by the various designers. However, it is practical to review and improve the concept design prior to the planning permission application, often regarded as a non-return gate. The concept design should include:

- A statement of the design direction
- The site layout and access, identifying ground conditions and planning constraints
- A detailed cost plan and schedule of activities
- Dimensioned plans, elevations and sections
- An outline specification for environmental systems
- The risks and a risk management strategy
- The procurement plan
- The project execution plan with key milestones
- Performance measures

The concept design workshop is likely to be of one to two days (8 to 16 hours) duration and involve between 7 to 12 participants comprising the design team and the client project sponsor/manager. Because the client value system is defined in the brief, it is no longer necessary to involve members of the client's board or senior management. Workshop tasks include:

- Presentation and team building
- Review function logic diagram and concept brief, particularly functional space requirements and adjacency
- Allocate costs from the cost plan to functions and spaces
- Highlight mismatches and risks
- Brainstorming solutions
- Evaluation and development of solutions
- Presentations
- Plan for implementation of preferred solutions
- Prepare action plan
- Sign-off by participants
- Facilitator to prepare list of solutions and action plan as a report

Detailed design workshop – final sketch design (FSD)

Once the client project manager has 'signed-off' the concept design the project team should begin the development of the detailed design and specification of the performance requirements for elements of the facility. The sketch design should freeze as much of the design as possible, defining and detailing every component of the construction work. It should identify risks associated with the project and outline proposed action if they arise, assess the quality requirements and define how success will be measured. The detail design should include:

- A statement of scheme design
- Location of facility and infrastructure requirements, planning approvals and other detailed permissions agreed
- Dimensions of spaces and elements provided with some description of components
- Performance specifications for environmental systems and services
- The cost plan
- Proposals for the maintenance and management of the completed facility
- The project execution plan with key milestones and targets
- Performance measures

The detailed design workshop is likely to be of two to three days duration and involve between 7 to 12 participants comprising the design team, the client project sponsor/manager, and the client facilities manager. Workshop tasks include:

- Presentation and team building
- Review function logic diagram, concept brief and outline sketch design
- Identify element and component functions
- Allocate costs from the cost plan to element and component functions
- Undertake life cycle costing studies
- Highlight mismatches and risks
- Brainstorming solutions
- Evaluation and development of solutions
- Presentations
- Plan for implementation of preferred solutions
- Prepare action plan
- Sign-off by participants
- Facilitator to prepare list of solutions and action plan as a report

Work in prime contracting has highlighted the benefits of organising work into clusters. This useful approach has been extended to represent the BCIS (Building Cost Information Service) standard elements into functional clusters as shown in Fig. 5.6. The logic of this approach is that it best reflects the client's value system through the work of the specialists comprising the cluster. For example, the

Cluster	Cluster function	BCIS element
STRUCTURE	Support load Transfer load	Substructure Frame and upper floors
EXTERNAL ENVELOPE and WEATHER SHIELD	Protect space Express aesthetics	Roof External walls Windows and external doors
FUNCTIONAL SPACE, CLIENT ORGANISATION and STYLE	Serve client function	Stairs Internal walls and partitions Internal doors Wall finishes Floor finishes Ceiling finishes Sanitary appliances and plumbing Communications and data
INTERNAL ENVIRONMENT	Maintain comfort	Space heating Ventilation and A/E Electrical
INTERNAL TRANSPORTATION	Minimise walking Raise/lower people and loads	Lift and escalator
EXTERNAL INFASTRUCTURE	Protect property Circulation/parking Remove waste	Site works and drainage External services

Project mission and function of building {

Fig. 5.6   Functional cluster groups.

client's value system may require a particular approach to aesthetics or to the quality of internal fit-out. The former is reflected through the external envelope cluster and the latter through internal space, quality and organisation.

Implementation workshop

The aim of this workshop is to ensure that recommendations set out are being implemented. The workshop will follow the concept and detailed design workshops. During benchmarking, examples of high implementation rates were cited where an implementation workshop was held. An indicative duration was given at between a half and one day. The potential participants were as in the workshop plus a representative of senior management. The latter added to the authority of the exercise. Workshop tasks include:

- Recommendations and findings of the workshop being followed up and summarised in a presentation by the participants.
- Workshop participants make presentations on progress within their allotted duties from the workshop action plan. Implementation issues are included together with problems encountered.
- Brainstorming solutions to problems encountered or recommendations that have run into difficulties. These are reviewed and refined.
- Brainstorming of new implementation issues, together with reviewing and refining.
- Recommendations and a final action plan are developed.
- Sign-off by participants.

5.8   Example

This example is based upon a fictitious project and describes function logic diagramming. The other techniques outlined above are not described.

A local authority owns a country park. The park comprises a mixture of woodland, open grassland, meadowland and a water meadow within the floodplain of a river. There is no bridge across the river, which divides the park into two. There is one main entrance to each part of the park.

The local authority, prompted by local schools, has decided to build a 'Park Centre' comprising a café, space for an educational exhibition and a permanent guide to the natural environment of the park. There is also to be a lecture/meeting room to accommodate approximately 40 people. A footbridge is to join the two sections of the park.

An early design showed the Park Centre located at one of the entrances (to minimise the installation of services) and a suspension bridge joining two high points in the park upstream of the meadow. Opposition to the scheme has come from the schools and residents on the side not served by the Park Centre. The cost of the suspension bridge has considerably exceeded the available budget. It has been argued that the bridge cannot economically be sited on the meadow due to the problem of flooding. In this context it was decided to hold a value management meeting.

Stage 1: Pre-workshop interviews

One of the two facilitators appointed for the project conducted interviews to determine the views of various stakeholders, identified as: residents; parents; schoolteachers; councillors; and representatives of the local authority administration, principally architectural services, accounts, leisure and recreation, and education. A group of 16 stakeholders were invited to a value management workshop.

Stage 2: Issues analysis

An issues analysis was conducted addressing such topics as:

- Community
- Education
- Health
- Politics
- Budget
- Location of the project
- Time
- Contractual issues
- Staffing and maintenance

The issues analysis ensured that all who were present at the workshop had all of the information impacting the project. A sorting exercise on 'what is driving this project?' highlighted the importance of education and health and also the restricted budget. The project was seen to have failed if: (1) the Park Centre was not equidistant from the two entrances; and (2) a means of crossing the river was unavailable.

Stage 3: Function analysis

The facilitators' task, once all of the information is available, is to focus the team on the prime function of the project. The focusing technique begins with a brainstorming exercise concerned with 'what exactly are we trying to do?'.

The team brainstorm functions that are required by the project. These functions may be high-order executive functions or relatively low-order wants. All functions are expressed as an active verb plus a descriptive noun and are recorded on sticky notes and scattered randomly across a large sheet of paper.

At the completion of the brainstorming session the team are invited to sort the sticky notes into a more organised form by putting the highest-order needs into the top left-hand corner of the paper and the lowest-order wants into the bottom right-hand corner. All other functional descriptions are fixed between these two extremes. It should be emphasised to the team that this is an iterative process and therefore any team member is entitled to move a previously ordered sticky note. Although this sounds confrontational it is very rare for disagreement to occur and ultimately the correct ordering of all of the functions is achieved.

How the next phase is undertaken is the subject of some debate. Some facilitators will allow the team to construct the diagram in its entirety. However, this can be difficult where there is a large team. The preferences are for the facilitator and perhaps one or two team members to delay going to lunch and to complete the diagram. Later, discussion can take place on whether the diagrams and particularly the prime function is correct.

Generation of functions

The functions were generated by the workshop team in a random fashion on sticky notes and placed on a large sheet of paper (see Fig. 5.7). It is the structuring of the diagram that prepares the team for innovation. The arrangement of the sticky notes for the workshop is shown in Fig. 5.8. In a very general sense the logic of the diagram answers the question 'how?' when working from left to right and 'why?' when working from right to left. Highest-order needs are placed at the top of the diagram and the lowest-order wants at the bottom of the diagram. A scope line divides the primary function of

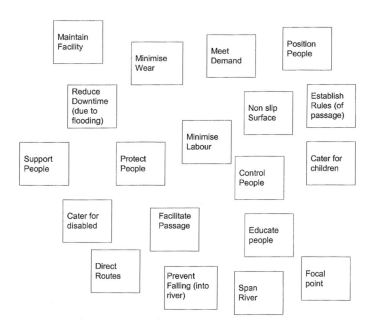

Fig. 5.7   Generation of functions.

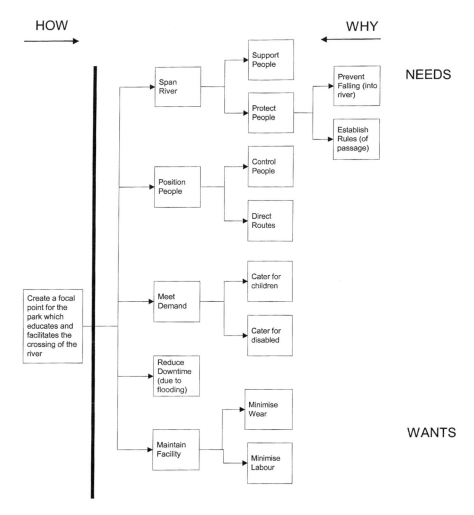

Fig. 5.8   A function logic diagram.

the project from the functions that may form the basis of brainstorming.

Innovation

The innovation stage involves brainstorming solutions; in this case they may include:

(1)   Stepping stones
(2)   Bridge
(3)   Concrete pipes with a robust surface to withstand the power of flooding

(4) Ferry
(5) Aerial runway
(6) Rope swing

A review of the above against the function logic diagram will prove which are the better solutions. Ideas (2), (3) and (4) may be carried forward for further development.

Further development within the team may reveal the following.

- In the case of this particular project the final solution might not be a bridge but an old-fashioned cable-stayed ferry boat, operated by a full-time ferryman, which uses the power of the river acting on the rudder to move the ferry across the river. The ferry could be accessed from a boardwalk constructed to withstand the flooding. The Park Centre could be at the head of the boardwalk on the nearest firm ground above flood level.
- Alternatively, a simple low bridge, which could withstand the force of flooding, might be a better solution.
- Or finally, large concrete pipes in the river may provide a deck for a walkway that would flood occasionally.

The above ideas would be taken away for further development and life-cycle cost analysis outside of the workshop. The final solution would be that which was within budget and met the quality requirements of all the stakeholders. The generation of the function logic diagram by a team is a powerful way of reaching consensus on the primary function and preparing the way for brainstorming.

5.9 Conclusion

This chapter has introduced the topic of value management, described a research project concerned with the international benchmarking of value management practice, described characteristic value management activity at various stages of a construction project and given an example of the use of one of the techniques commonly associated with the value management toolbox. Value management has reached a level of maturity within manufacturing and construction whereby the style and content of the various workshops is reasonably predictable. Many of the tools and techniques of value management can be transferred to other workshops, for example bidding conferences, partnering workshops, etc. However, considerable further research work is required into methods of measurement of the client's value system expressed in a form suitable for auditing. This work would considerably advance the use of value management in the public sector where an auditable approach to projects is vital.

References

BRE (1997) *Value from Construction; Getting Started in Value Management.* Building Research Establishment.

CIB (1997) *Briefing the Team.* Thomas Telford.

CIRIA (1995) *A Client's Guide to Value Management in Construction.* Construction Industry Research and Information Association.

European Commission (1995) *Value Management Handbook.*

HM Treasury *CUP Construction Guidance No. 2 Value for Money.* http://www.hm-treasury.gov.uk

ICE (1996) *Creating Value in Engineering.* Thomas Telford.

Kelly, J. & Male, S. (1993) *Value Management in Design and Construction: The Economic Management of Projects.* E & FN Spon.

Male, S. & Kelly, J. (1998) *The Value Management Benchmark: A Good Practice Framework for Clients and Practitioners.* Thomas Telford.

Norton, B. & McElligott, W. (1995) *Value Management in Construction: a practical guide.* Macmillan.

Shillito, M.L. & DeMarle, D.J. (1992) *Value: its Measurement, Design and Management.* Wiley.

6 Risk Management

Nigel Smith

6.1 Introduction

Risk and uncertainty are inherent in construction. Risk management is the term given to the overall process of managing construction under conditions affected by risk and uncertainty. The terms 'risk' and 'uncertainty' have different meanings, in contradiction to some publications where the terms are used interchangeably. 'Uncertainty' is used to reflect a genuine unknown that could have positive or negative effects on a project, but risk can be estimated and has a negative effect only. The understanding of risk management is made difficult by two factors: firstly, a lack of clarity of the purpose of risk management; and secondly, that risk management is an iterative process reflecting the dynamic nature of risk within the project life cycle. The purpose of risk management is to provide information to empower the project manager to make a better decision, at any time during the project life cycle, with respect to the project itself. Risk management is not about prediction. If any risk analyst could use risk management techniques to accurately predict the future then that person would be extremely wealthy, sought-after by the worlds' leading industrialists and politicians, and regarded as the modern-day equivalent of the Oracle at Delphi.

The term risk management is used by different industrial sectors to describe discrete activities that not only occur at different points on the project life cycle but are cyclic or repetitive processes involving different levels of certitude and possibly different methodologies. It is therefore not surprising that a degree of confusion exists. In order to clarify the risk management process and remove some of these difficulties, this chapter will consider the use of risk management throughout the project management cycle, before concentrating on the construction phase in more detail.

6.2   Risk management over the project life cycle

Risk in this context can be considered as the 'whole project risk'. That is to say that the risk management has to recognise boundaries within which it operates; for example, corporate business risk is not usually included in a project risk analysis. Equally it must be remembered that the client must work in terms of project life whereas contractors are often only able to work in terms of contract life. A schematic view of the relationships between project cycle, cashflow and contract type is given in Fig. 6.1. The immediate left-hand side of the project life cycle, the earliest stages of a project, are frequently named as the 'conception' or 'initiation' stage. At this point the idea of a project has resulted from a process of strategic management. The first task for the risk management process is to inform the project sponsor or project manager who is faced with the decision as to whether the project is viable or not. This is an important but difficult decision due to the high levels of uncertainty surrounding the project at this stage.

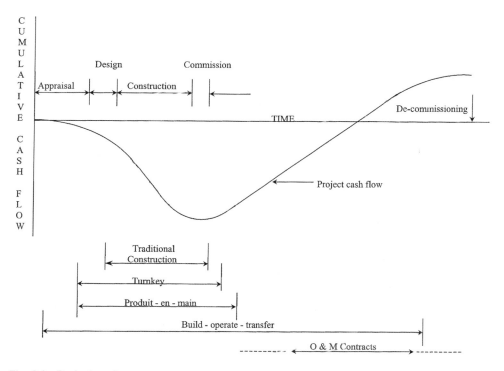

Fig. 6.1   Project cycle.

Project viability

There are many more project ideas than become viable projects and the early identification of a poor investment option could result in saving time, costs and resources. However, the incorrect rejection of a profitable project could result in financial problems for the client. There are a number of methods of risk management utilised at this stage; typically, a checklist or hurdle method would be adopted as there are normally insufficient data available to perform a formal risk analysis.

Checklist methods consist of considering the implications for the particular project of various external factors, often awarding a score for each factor. Projects with a total beyond a certain value or hurdle would be discontinued. However, care needs to be taken with the choice of metric and its assessment; for example, if 'political stability' were the factor and 'number of changes in government' were the metric, Italy, with more than 50 different administrations since 1945, would probably score very highly but no one would suggest that a project in Italy should be abandoned because of political risk. Other values are often financial, frequently based on a given percentage internal rate of return, (IRR). Comparison would again be made with a hurdle such as minimum lending rate or a pre-determined value of IRR set as a minimum for projects in that industrial sector. Projects not abandoned at viability progress to the feasibility study stage.

Feasibility stage

The formal process of risk management is to assist in the identification of risk sources, the quantification of their potential effects and guidance on mitigation and management in order for the project manager to make a better decision. In the UK, although there have been a number of recent risk management guides – from the Association for Project Management, the Institution of Civil Engineers, The Chartered Institute of Building and Faculty of Actuaries, the Construction Industry Research and Information Association, and the HM Treasury Procurement Group (formerly Central Unit on Procurement) – they all have the common philosophy of the processes of identification, assessment and management and all utilise a money value for comparing risks, normally IRR.

Identification

In a process directly comparable with value management, the identification of risk sources takes place as the first stage in the process. It is important to note that it is the sources of risks or rather the potential sources of risk that are being sought at this stage. Risk sources are identified in three main ways: historical projects, industrial checklists and workshop brainstorming sessions. Historical projects can be misleading because the project may not be similar enough, or not in a similar location or with similar objectives, and even where these elements are true the information may be based on outdated techniques and processes that no longer apply. Industrial checklists are published in some risk guides or by relevant authorities, in particular industrial sectors such as the offshore oil operators. The lists serve as reminders to force the risk analyst to consider the potential sources of risk from the proposed project. The brainstorming sessions are exactly equivalent to the value management approach but the object is to identify as many potential sources of risk as possible.

After one or all of the identification methods have been conducted the analyst is left with a long list of potential sources of risk. The analyst then classifies the risk sources on the basis of those sources with a high probability of occurrence and a high impact on the project. It is only those risk sources, if any, which can be classified both as 'high impact' and 'high probability' that are of real interest to the risk analyst.

Assessment process

The basic process underlying the previously mentioned standard guides consists of two further stages: assessment of risk and management of risk. As with the first stage of identification, there is no definitive approach but rather an iterative approach whereby projects that are more risky require more detailed work. If, after identification, no high-impact and high-probability risks have been found then a simple 'hurdle' assessment of risk – such as 'is the return greater than the minimum bank lending rate?' – might be sufficient for the project.

In numerous construction industry projects, high-impact and high-probability risk sources will have been identified and therefore the evaluation of their effects is undertaken. Usually the first step is to produce a sensitivity analysis diagram, sometimes referred to as a spider diagram. The diagram is a graph with axes representing the project performance (usually in financial terms – profit, NPV or IRR) and percentage change in a key variable. Each key risk is taken in turn, altered by a given +/− percentage and the resultant project value plotted. The methodology relies upon two assumptions: firstly, that each variable is independent; and secondly, that each variable can change by any percentage without changing the project.

Neither of these assumptions is likely to hold for all the key risks on a project, nevertheless the advantages of distinguishing the most sensitive risks quickly, outweighs the possible inaccuracies.

A typical sensitivity diagram is shown in Fig. 6.2. Notice that not all the plotted lines for the variables are straight – some are curved and some variable. For some variables, change can be positive and negative, whereas for others the results are in one sector only. The way to interpret the diagram is to identify the key variable plots closest to the project axis. These are the most sensitive variables as the smallest percentage change in the variable has the relatively largest effect on the project parameter. Similarly, the key variable plots closest to the change in variable axis are the least sensitive.

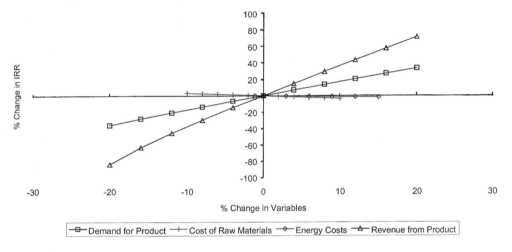

Fig. 6.2   Sensitivity analysis diagram.

It is possible to enhance the value of a sensitivity diagram if the analyst is experienced enough to add one or more 'probability contours'. In the same manner in which a contour line on a map joins points of equal height, a 'probability contour' joins points on the key variable plots with an estimated equal probability of occurrence. There is no precise or correct method for calculating a probability contour and it should only be attempted by experienced analysts with a deep understanding of the construction processes being assessed. In terms of communicating information to other members of the project team, probability contours are particularly useful in showing the likely amount of change, i.e. there might be a 50% chance that variable 1 will increase by 5% whilst there could be a 50% chance that variable 2 will increase by 200%. It is also useful in identifying to what extent variables are likely to increase or decrease. With some variables the probabilities could be about equal; with others they may be far more likely to increase than decrease. Whilst helpful, probability contours are indicative rather than absolute and unless the analyst is confident that there is sufficient evidence to support the drawing of a contour they should be omitted.

It is possible that there would be no sensitive risks identified and the assessment process could end. However, in most cases one or more sensitive risks will have been found and a more detailed assessment is required. For the next level of evaluation, risk management software (RMS) is essential. It is likely that RMS would be used to generate the sensitivity diagrams, although it is possible to produce them manually or by using a macro on a standard spreadsheet package. There is currently a wide range of commercial RMS packages designed to run on PCs without requiring any unusual specifications. The majority of the commercial RMS packages adopt the Monte-Carlo simulation technique to produce a probabilistic analysis for the project.

There are many textbooks describing Monte-Carlo simulation but here only a brief outline of the key points will be given. A model of the project, usually based upon a critical path network, is produced and sufficient data to allow a time and cost performance of the project to be calculated. Only a small number of the model activities will represent sensitive variables and it is for these activities only that a 'triple estimate' of activity duration is needed. However, unlike the sensitivity analysis in this case all the variables are considered simultaneously. A 'triple estimate' is a practicable means of obtaining a distribution for the likely duration of the activity. It is called triple because it requires three values to be determined: the most likely value, the shortest value (or optimistic value) and the longest value (or pessimistic value). It is unlikely that the distribution will be symmetrical, as in practice the shortest time and the most likely time are usually close, but if things go wrong the longest time could be substantial. After a triple estimate has been obtained for all of the sensitive variable activities, the computer converts these values into a unit duration and calculates the probabilities for each time period within the distribution. Next comes the process which gives the technique its name. The package will generate a random number, R, within the unit range, $(0 < R < 1)$, for each sensitive variable activity. The value of R is then converted into a duration for each of the sensitive variable activities and the model is used to calculate a project parameter; usually IRR. The first iteration is then repeated normally between 2500 and 10 000 times. A cumulative frequency plot of iteration against IRR can then be plotted, as shown in Fig. 6.3.

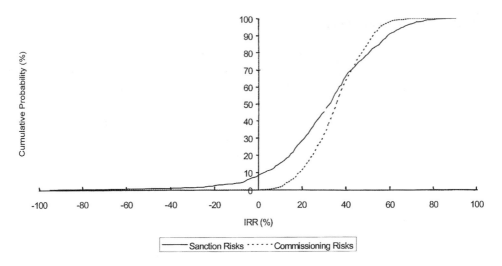

Fig. 6.3   Cumulative frequency plot.

The cumulative frequency plot or S-curve is a representation of the profitability and the risk of that particular project option. It is common practice to consider the curve between the 15 percentile and the 85 percentile lines, thus removing the boundary effects, and indicate a range of likely IRR values obtainable from the project. The narrower the range, the less risk is associated with the project. However, the best use of the S-curve is in comparison with other curves and a number of project options should be compared and the most appropriate option selected – which may include the abandoning of the project entirely.

With a known project option the management phase can commence. Risks can be avoided, transferred or retained. The sensitive risks should be reviewed to ascertain whether it would be practicable to make changes to the type, process, material or resource that might mitigate the risk. Sometimes this is possible but often through value management or other procedures there are no further changes possible to avoid risk. Risks can be transferred to another party, typically through contract or surety or bonding; however, the fundamental principle of risk transfer is that the party best able to control, manage or sustain the risk should be the party with responsibility for the risk. If a risk is transferred that is contrary to this principle then in order to remain in business there must be a price connected with the additional risk. That price would be paid irrespective of whether the risk event actually happens.

Residual risk management

Finally, the residual risk has to be managed, and increasingly risk registers and risk management plans are being used to formally monitor and control this process. These processes will be discussed in more detailed in the next section.

It would be wrong to give the impression that this process of risk management, of identification, assessment and residual risk ownership was a single cycle undertaken at one specific point during the feasibility stage. Risk management is a continuous and dynamic process which is repeated frequently throughout the feasibility stage. It is repeated to test different ways of modelling risk and to reflect the increasing amount of certainty concerning aspects of the project as development proceeds. In this way the risk model is as up to date and as helpful for decision-making as possible. One of the key decisions for which information about risk is essential is the choice of appropriate contract strategy.

The end of the feasibility stage is marked by the decision to either sanction the project or to abandon it. During the feasibility stage it is usually the client (owner) or the project manager who has been conducting the analysis of risk. As the trend towards types of collaborative working increases, more construction professionals are finding themselves involved in this early pre-tender risk management stage, often as part of a partnership alliance of a PFI promoter group.

6.3   Design and construction

The construction project will usually go out to tender after sanction. Using traditional UK procurement routes this may be for construction only, although design-and-build is a popular option for many types of construction work. This is often the first opportunity for the contractor's staff to undertake risk management. For any major or complex project the risk management process should be undertaken, particularly for design-and-built projects with a constructability interface and for PFI-type projects where raising finance and managing risk are important.

The process would adopt the same steps as described above for the client's staff, namely identification, assessment and residual risk management. It is as important for the contractor to be aware of risk as it is for the client. For a contractor, the risk management process is in two stages: the first stage is to identify risks which could seriously damage the future prospects of the company and which might cause the contractor to decline to participate or tender; and the second stage is to manage the risks effectively to increase competitiveness and profitability. The knowledge obtained from a risk management approach can assist with preparing the tender, selecting the resources and method of construction, interpreting the contract and pricing the works.

The aims of risk management are firstly to ensure that only projects that are genuinely worthwhile are approved and secondly to avoid excessive overruns which can invalidate the economic case. Risk management is one of the most creative tasks in project management. In terms of construction project management it achieves this through generating realism and thereby increasing the commitment to control. One of the most popular vehicles for this is the risk register. This is the stage that is most familiar to most construction professionals. The register identifies each of the risks; the type of risk – for example, is it a hard, technical risk or a soft process risk?; and the party with the ownership of the risk. There is no standard form for a risk register but some contain further details of the risk such as the extent to which float, contingencies or tolerances have been utilised in the contract to mitigate the consequences of the risk, together with further details of alternative risk mitigation measures. Some clients have requested that a risk register be submitted with other tender bid documents.

The register is also used as a live document during the works. It should be updated and modified as time progresses, reflecting the occurrence or non-occurrence of identified risks and the dynamics of any new or unidentified risks that occur. The register concentrates on the remaining, future part of the project still to be completed and becomes a shorter and shorter document as completion and/or commissioning is approached. The register is a valuable project management tool and is an important element in any decision-making process concerning the contract.

In order to manage a project it is necessary to apply monitoring and control procedures. At periodic intervals, or when a major change occurs, it is necessary to repeat the identification, qualification and response cycle and subsequently update the risk register. This process is also necessary to reflect the progress of the project. As time passes, some risks will be removed from the analysis as they failed to occur, whereas others, actual or theoretical, might need to be incorporated. At all times, the project sponsor or project manager requires an updated assessment on which to base project management decisions.

6.4   Operation and maintenance

For many projects the operation and maintenance phase is either transferred to another party or is known to be low risk and suitable for routine preventive maintenance techniques. However, for some clients and for some contractors engaged on operation-maintenance and training-type contracts, there may be significant risk to be managed. Projects such as the servicing of a bascule toll bridge, or the repair of broken rails on a major permanent way rail network or the shutdown and refurbishment of a chemical refinery all have significant safety, financial and technical risk to be managed.

The party contractually responsible for the maintenance work would undertake a specific analysis for the particular operation. Despite the fact that there may be previous records of similar operations to assist with the identification stage, it is recommended that the complete risk management process is followed for each specific event. The output of this process will again be in the form of a risk register, albeit over a shorter time period. The same process might be required if there is a decommissioning stage to the project.

6.5   The case study

Background

The project case study, based on a contractor's tender for a military barracks in the UK, provides a useful insight into a contractor's role in tender risk management. The project is undertaken using the PRIME contract system adopted by Defence Estates in the UK, which places special emphasis on the identification and effective management of risk.

As is becoming common practice in the UK, the project illustrated the difficulties of defining boundaries between value management and risk management. Whilst seemingly having much in common, the basic philosophies of risk and value management are different. Value management is essentially concerned with maximising opportunities and eliminating waste whereas risk management is about making decisions under conditions of uncertainty to avoid or mitigate potential adverse consequences. Hence the two procedures required different mind-sets from the participants to the process.

PRIME contracting requires both the maximisation of opportunities and effective risk management. The case study indicates this by describing a facilitated workshop for an issues analysis and a function diagram where the importance of risk ownership and risk transfer and acceptance is acknowledged. A risk analysis was then undertaken leading to a listing of general risks, which were then arranged under the headings of the four clusters identified for the project: (1) envelope and shell, (2) demolition and infrastructure, (3) mechanical and electrical, (4) finishes and compliance and maintenance risks. The risks are identified as specific technical risks or process risks, with other risks usually covered in the conditions of contract being non-attributed. The next stage would be to allocate the risk rating by assessing its impact and likelihood of occurrence. The risks scoring a high value in both categories would be placed on the cluster list and an appropriate owner from the cluster allocated to the risk using the fundamental principle of responsibility linked to control and management. The risk owner then has to prepare a plan of how the risk is to be managed and this is entered into the risk register. The register should be updated every month.

The purpose of the project is to provide sleeping and living accommodation for service personnel, plus parking and related facilities to the standards and constraints indicated in the tender documentation.

There is a requirement for mixed facilities to be provided for living-out personnel to a standard comparable to those in the JSP Scales. The facilities will include cloakrooms, male/female lockers, changing areas and showers. It is anticipated that the number of personnel to be accommodated is in the region of 200.

Issues identification and prioritisation

A facilitated workshop was held with the prime contracting team which commenced with an issues analysis using the following process:

- Brainstorming of any issues of concern using sticky notes and flipcharts
- Group and theme

The following ten issues emerged as the most important:

(1) Roles and responsibilities
(2) Whole-life costing in design
(3) Who owns the various risks?
(4) What level of design is required for the indicative cost plan?
(5) Design and cluster integration with associated communication
(6) A clear definition of the clients value system required
(7) What are the client's needs?
(8) Integration of the supply chain into design

  (9)   Identification of appropriate clusters
  (10)   What is the whole-life period?

Function analysis

As part of the issues analysis, the team constructed a function diagram to establish, by consensus, the main purpose of the project. The resulting diagram is given in Fig. 6.4.

Risk analysis

A risk analysis was carried out by the team during the first part of the second day to establish technical and process risks associated

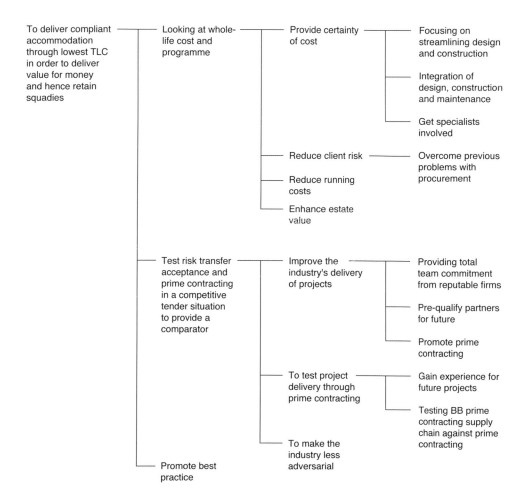

Fig. 6.4   Project function diagram.

Table 6.1   Details of risk analysis.

Risk description	Risk type
1. Prototype designs	
2. Design meeting cost plan	Process
3. Inflation	Technical
4. New process	Process
5. Security	Technical
6. Military activity	Technical
7. Adequacy of cost plan	
8. Contract conditions	
9. Project sponsor	
10. Resolving contractual links	Process
11. Cluster coordination	Process
12. Location and resources	Technical
13. Time/phasing	Process
14. Client change	Technical
15. Sequence continuity	
16. Failure of cluster	Process
17. Cash flow	
18. Health and safety	Technical
19. Communication	Process
20. Access (logistics)	Technical
21. Environmental impact	Technical
22. Client culture	Process
23. MoD supply chain	
24. Industrial relations	Technical
25. Supply chain culture	Process
26. Open book	Process
27. Fuel contamination	Technical
28. Weight limits on routes	Technical
29. Restricted height on access	Technical
30. Decanting – vacating on time and possessions	
31. Relationship with residents	
32. Weather	
33. Interfacing with third parties	Process
34. Changes to the brief	
35. Client approvals procedure	Process
36. Changes to legislation	
37. Specialist advisors' approvals	Process
38. Vandalism and theft	
39. Fit for purpose	
40. Client suppliers	
41. Contract termination	
42. Antennae	Technical
43. Siting board	Technical
44. Market conditions – costs and inflation	
45. Redesign based on supply chain ideas	Technical
46. Liquidation insolvency of member of the supply chain	
47. Unique supplier raises prices	
48. Innovative design	Technical

Contd.

Table 6.1 *Contd.*

Risk description	Risk type
49. Prefabrication tolerance and fit	Technical
50. Latent defects	Technical
51. Quality of work	Technical
52. Product suitability	Technical
53. Risk of terrorists	
54. Risk of demonstrators	
55. Change of government policy	
56. Archaeological	
57. Ecological	
Envelope and shell risks:	
58. Availability of materials	Technical
59. Labour availability	Technical
60. Leaks	Technical
61. Airtight	Technical
Demolition and infrastructure risks:	
62. Ground conditions and contaminants	Technical
63. Scope of infrastructure work	Technical
64. Asbestos	Technical
65. Bomb shelter	Technical
66. Hidden munitions	Technical
Mechanical and electrical	
67. Interface with existing infrastructure	Technical
68. Failure to comply	Technical
69. Commissioning	
70. O&Ms	
71. Builder's work in connection	
72. Maintaining existing services	Technical
73. Service capacity	Technical
74. Unknown services	Technical
75. Disconnection of infrastructure services – temporary supply	
76. Damage of existing services through connection	Technical
Finishes risks:	
77. Acceptable standard	Technical
78. Durability	Technical
Compliance and maintenance risks:	
79. Failure to comply	Technical
80. Durability	Technical
81. Scope of compliance	Technical
82. Conflict between designers and facilities management	Process
83. Through-life cost input to design	
84. Proving period	
85. Hand-over procedure for phasing	Process
86. Interpretation of compliance	

with the four working clusters identified previously. The four clusters were:

(1) Envelope and infrastructure – single cluster leader
   • Envelope and shell
   • Demolition and infrastructure
(2) Mechanical and electrical
(3) Finishes
(4) Compliance and maintenance

The resulting analysis is presented in Table 6.1.

These risks will be analysed further with specialists being assigned to manage in an appropriate manner, i.e. the best expertise for the job within each cluster.

Structure underpinning the partnering agreement

The relationships between the risk analyst in the project management team, the prime contractor, the core team and the cluster leaders were established and mapped out as shown in Fig. 6.5.

6.6 Risk management guidance

There is a wealth of material available to construction professionals and others providing guidance on the management of risk. This

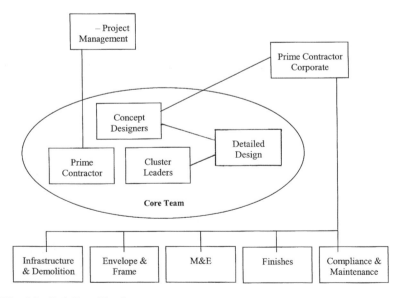

Fig. 6.5 Relationship diagram.

chapter has dealt largely with the underlying principles of risk management and the significance of risk management in construction. Recommended publications are given in the Further reading section at the back of the book.

6.7   Conclusion

In the early days of risk management the procedures were not commonly agreed, the supporting software was severely limited in its ability to reflect actual practice and few professionals had received adequate risk management training. Thankfully in the UK and in many other countries, the situation is now much improved. Published standards and codes of practice have a common basis, the supporting software is more powerful and more user-friendly, and most first engineering and technical degrees contain modules dealing with risk and uncertainty.

7 Building Project Price Forecasting

Chris Fortune and David Weight

7.1 Introduction

This chapter is concerned with the theory and practice of forecasting building project prices for clients. Such forecasts need to be formulated by construction industry professionals before and during the early design stage of building projects in order to ensure clients can make strategic decisions and obtain value for money. Conventionally, in the UK, it is the responsibility of clients' professional cost consultants to provide building project price advice. Such a professional service needs to ensure that effective communication has taken place between advisor and client in terms of the early project price advice provided and its impact on the project's anticipated risks.

The chapter begins by examining current theories in relation to the processes by which building project price forecasts are compiled. The chapter then reports current practices and establishes which of the available types of model or technique are in actual, as opposed to theoretical, use. The chapter indicates a framework of overarching issues that practitioners need to address in order to benchmark their model selection decisions. The chapter goes on to identify emergent issues related to the practice of building project price forecasting and concludes by providing an illustration of a knowledge-based computerised price modelling system that will be typical of the tools used by practitioners to formulate building project price forecasts in the future. Construction industry professionals who are presently situated in organisations that use such knowledge-based systems are currently at the 'leading edge' in terms of the actual practice of building project price forecasting. Other organisations that aspire to providing a truly professional building project price forecasting service to their clients will, in the future, need to develop similar in-house expertise.

7.2   The process of building project price forecasting

Current building project price forecasting

A feature of the construction industry is the need for its prospective clients to be advised of their likely financial commitments prior to their commissioning expensive project design work. There exists a popular perception amongst clients, academics and other design professionals that the provision of early-stage strategic financial advice has often been of poor quality. The Egan Report (1998) called into question the quality of building project price advice when it commented, in relation to underachievement in the industry, that, 'more than one third of major clients were dissatisfied with their consultants ... in providing value for money'. The provision of reliable early-stage building project price advice is a key element by which clients can ascertain the extent of financial risk as well as whether they are obtaining value for money from the organisations they employ to deliver their building needs.

Many academics have long expressed dissatisfaction with the way practitioners approach the provision of early-stage building project price advice. They have in the past called for practitioners to make a 'paradigm shift' away from the selection of traditional manual single-point deterministic forecasting methods. It is accepted that such methods are incapable of reflecting the uncertainty and risks inherent in the building production process. In more recent times, Bowen and Edwards (1998), in their study into the quality of building project price advice, identified that the processes by which building project price forecasts are both produced and transmitted to clients are in need of continuing investigation.

To develop an understanding of how building project price advice may have its quality enhanced, one needs to appreciate the processes involved. To illustrate this it was decided to develop a picture of the advice function which shows its constituent parts in the context of accepted theory.

The complexity of building project price forecasting

Some of the processes involved in the provision of quality building project price advice have already been the subject of theoretical investigation. For instance, Bowen (1995) advanced a communications-based theory of price advice which identified that an interactive approach is needed in which communications and all key assumptions are transparent and understandable. Figure 7.1 models the complete building project price advice process from the reception of the project's brief to the decision of the client to proceed with

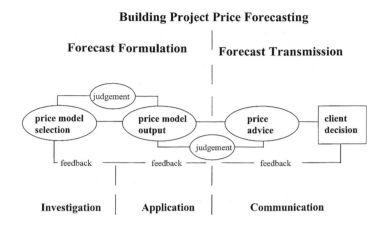

Fig. 7.1   The building project price forecasting process.

the project's design. The model reflects Bowen's (1995) communications-based theory by dividing the process into the two major
phases of forecast formulation and forecast transmission. The model
also indicates that the formulation phase of the process can itself be
divided into iterative cycles of model investigation and application.

Figure 7.1 shows a complex set of interrelated activities that
involve practitioners making decisions and exercising judgement at
differing points throughout the process. The main activities call
upon practitioners to select and use an appropriate price forecasting
model (or models), consider its output and communicate a judgement to clients on the anticipated building project price. Practitioners need to capture and utilise feedback on the judgements they
make within the investigation, application and communication
stages of the process and not just feedback in terms of the accuracy
of their final forecast, if they wish to enhance the overall quality of
the service they provide.

In order to further develop theory and set out best practice
guidelines, academics have begun to study the actual, as opposed to
theoretical, human processes involved in formulating as well as
transmitting reliable early-stage building project price advice. For
instance, one such study was into the factors affecting model
selection, which is an activity within the investigation phase of the
building project price advice process (illustrated in Fig. 7.1). As a
result of that work, which developed a constraints-based theory of
model selection, it has been shown that practitioners involved in
selecting a building project price forecasting model are involved in
making judgements on a complex range of factors. It has been found
that the process of selecting building project price forecasting
models is an iterative process that is affected by factors that can be

categorised as being related to (1) the project forecaster becoming aware of the project and its needs, (2) the evaluation of the response level required by the client, and (3) the assessment of the resources available, within the overarching constraints of the practitioner's organisational setting and own level of model understanding.

Factors affecting building project price forecasting

Figure 7.2 provides an illustration of the factors that have been found to impact on the model selection process and indicates typical issues that practitioners need to address in order to select the most appropriate model(s) for use. On becoming aware of the requirement to produce an early-stage building project price, forecast practitioners need to address issues connected with client sophistication, project typicality, time availability, and the nature of the drawn and non-drawn data available. Similarly, practitioners, when assessing the level of response required, in terms of selecting the most appropriate type of building project price forecasting model, need to consider issues such as the basis of the service commission, the organisational approach to service response and the nature of the output from each of the types of model under consideration for selection. Final selection of the most appropriate type of forecasting model for the potential building project needs to consider issues related to the resources available to the practitioner, such as:

- The extent of any available cost database.
- The nature and location of any organisational support facilities.

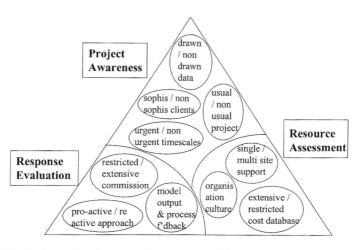

Fig. 7.2   Factors affecting the selection of building project price forecasting models.

- The prevailing organisational culture towards the selection and deployment of particular types of price forecasting models.

Further work that addresses issues related to the role and impact of professional judgement on the building project price advice process would also help practitioners develop better quality forecasts for their clients. In order to further refine Fig. 7.2 and identify the actual factors that influence practitioners in their choice of model types, it is firstly necessary to identify the models that are currently available and in actual use.

7.3   The identification of potential models in use

Fortune and Hinks (1998) conducted a nationwide survey with all of the 2300 cost consultancy organisations that were operating in England in 1998. The survey attracted a 64% response rate and established the state of the art in terms of the incidence-in-use of building project price forecasting models that were actually rather than theoretically selected for use. The results of that survey together with the results of an earlier regional survey are shown in Fig. 7.3. The left-hand side of Fig. 7.3 shows that the traditional manual types of price forecasting models are still in common use and the paradigm shift towards the newer non-traditional models has not yet been generally achieved. Of the newer computer-based models in actual use, only those related to the provision of life-cycle cost advice came close to being used by 50% of the survey's respondents. The techniques that had the least incidence of use were those in the statistical and knowledge-based categories such as regression analysis, time series models, Monte-Carlo analysis, and knowledge-based computerised systems such as ELSIE and other in-house systems.

The full statistical analysis of the survey's results revealed that there was no use made of the following models, namely the conference method, the comparison method, the cubic method, the storey-enclosure method, the graphical method, the superficial perimeter method, the exponent method, the financial method, dynamic and the linear programming method. In addition it was found that organisational type and size affected the incidence-in-use of non-traditional computer-based types of models. It was found that small-sized (fewer than five employees) organisations that styled themselves as being cost consultancy only (rather than multi-disciplinary or project management) types of organisations had lower frequencies-in-use of the non-traditional types of models. The survey also revealed that a major but not overwhelming barrier to

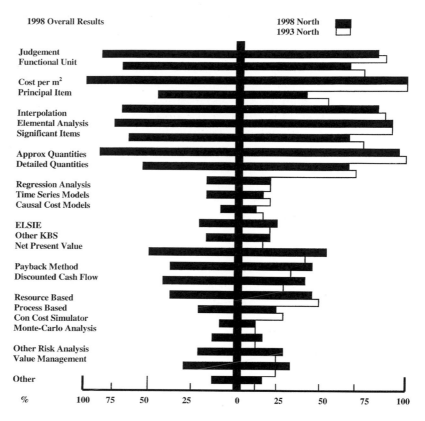

Fig. 7.3   Incidence of forecasting models in use.
Source: Fortune & Hinks (1998)

the use of some types of non-traditional computer based models was a lack of model understanding amongst practitioners.

7.4   Emergent issues

Following further manipulation of the survey data it was possible to directly compare the building project price forecasting models that were selected for use by respondents to both the 1998 national survey and a previous regional survey (Fortune & Lees, 1996) that was undertaken with organisations located in the north of England. The right-hand side of Fig. 7.3 reveals the results of that data manipulation and shows the incidence-in-use of the models that were selected for use by practitioners in those organisations in 1993 and 1998.

In general, the right-hand side of Fig. 7.3 shows that the traditional single-point deterministic building project price forecasting techniques had a reduced incidence-in-use, amongst the organisations indicated above, between 1993 and 1998. Figure 7.3 also shows that there was an increase in the use of the newer types of non-traditional models, such as life-cycle costs, knowledge-based expert systems, Monte-Carlo and risk analysis models, amongst the organisations surveyed in both 1993 and 1998. In addition it was found that cost consultancy organisations that classified themselves as being either a project management or multi-disciplinary type of organisation, had a higher incidence-in-use of the non-traditional more computerised building project price forecasting models. This evidence indicates that such organisations now constitute the leading edge of the profession in terms of building project price advice provision. Such organisations are in the process of redefining the paradigm of building project price forecasting and indicate that the shift called for in the building project price forecasting model selection process may, as yet, only be postponed amongst such multi-disciplinary and project management types of organisations.

The contextual variables of organisational type and size, together with lack of model understanding, act as a framework against which cost consultancy business organisations can benchmark the processes they use to provide clients with the building project price advice required. For instance, large-sized cost consultancy organisations wanting to improve the quality of the forecasts they produce for their clients would need to consider, (1) the models that they were selecting, together with (2) the selection factors found to be influential (see Fig. 7.2) and (3) the full range of models found to be used by other competitor organisations (see Fig. 7.3).

Hinks (1998) investigated the link between process capability and information management systems implementation in surveying organisations. He concluded that there existed an opportunity for organisations to develop a 'co-maturation' of the services they offered with the knowledge-based information management systems they utilised in terms of their organisation's capability to provide value-added services to their clients. Cost consultants in general may be said to have failed to embrace the potential for information technology to generate knowledge-based management systems capable of fundamentally redesigning their business processes. However, it is clear that there are some leading-edge organisations that are making full use of knowledge-based management systems to enhance their business performance. Practitioners concerned with ensuring that their clients receive full value for money and who are situated in organisations that do not measure up to the benchmarking framework indicated above, need to follow the example set by other leading-edge organisations within the profession. Such action will be required if they want to continue providing truly professional levels of building project price advice to their clients.

7.5   Case study – Live Options software

Currie and Brown Ltd, international cost and project management consultants, St Albans, UK, have developed a knowledge-based software package called Live Options which they use as a tool to provide their clients with early-stage building project price advice. The development, ownership and utilisation of such software tools will increasingly become necessary if competing cost consultancy organisations wish to provide their clients with similar levels of quality in terms of building project price advice.

The Live Options software package is an example of a knowledge-based software tool that has been developed in response to commercial requirements for reliable front-end building project price forecasting and value planning. The system defines and determines building project price forecasts not merely on the basis of historic cost analyses of past similar projects, but on the basis of the functional requirements and recognised design cost drivers of typical building projects. Such data have been assembled within the software package as a series of templates or cost models that professional cost consultants can use as the basis of building up bespoke project price forecasts. The system also acts as a checklist of all the major items involved in a project and can be used to carry out powerful iterative studies of alternative project designs and their associated costs.

At the core of the Live Options software package is an integrated suite of geometric and engineering programmes which provide interactive and real-time reports that can be interrogated fully by the client and the design team. The main components of the Live Options software package have been illustrated in flowchart form in Fig. 7.4. The capacity of the system to quickly present conceptual cost plans in an elemental format that is based on minimal briefing information at the initial stages of project consideration is a real step forward in the process of formulating building project price forecasts. The modelling system helps to assess the impact of changes in the client's brief in terms of its anticipated building form, performance criteria, aesthetic preferences and site constraints, and can provide meaningful and reliable comparative cost data which are presented in both a textual and graphical form. By generating such information, software tools such as Live Options allow cost consultancy practitioners to add value to a design team's processes. It can act as a catalyst for design decisions and facilitates under-

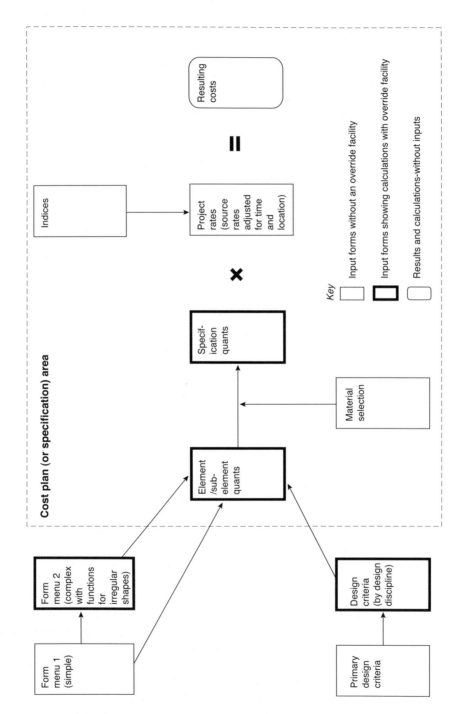

Fig. 7.4   The main components of the Live Options knowledge-based software tool.

standing and negotiation between members of the design team. This capability ensures a balanced and cost-effective construction response to the client's brief. In this way computerised knowledge-based tools can help practitioners save money for clients by instantly providing data for early decision-making, which in turn reduces wastage of resources. The explicit nature of the model is a particular feature of the system which avoids the consequent negative impact on resources of multiple cost/design iterations at later stages of the design process.

As the key cost drivers of a project are established and reviewed at the outset of the design process, Live Options allows the time at the front-end of a project to be better utilised by the design team to explore and develop the appropriate project concept. Typically, the conventional building project design process experiences problems related to (1) time-lags between the generation of potential design solutions and the assessment of their potential cost impact, and (2) the project approval costs not meeting the detailed requirements of a project. Such problems do not usually come to light until later on in the process and can often result in the project design team experiencing abortive time in either redesigning to meet the approved budget, or carrying out further exercises to reconfirm the client's original brief. The development and use of software tools such as Live Options establishes a communication and approval interface between client and design team through which function, design criteria and associated forecasts of project prices can be confirmed and explicitly determined. Such a facility addresses issues of effective forecast transmission and communication which were highlighted earlier in this chapter (see Fig. 7.1) as being critical in the provision of high-quality building project price advice.

In practice, clients are introduced to modelling systems such as Live Options at first meeting stage. At that meeting, data on a project's key design cost drivers, such as the proposed foundations, structure, coverings and fit-out features, are requested. If a client cannot supply such data, the system supplies a price based on experience. The data are then used by the Live Options system together with in-built data based on the design standards related to the contextual climatic and engineering consequences that flow from a project's initial briefing data. The model uses such data together with any already identified early-stage project component data to produce its project forecast.

The system comes into its own whenever any project data are revised as the impact of any proposed change is assessed across all aspects of a project's proposed design. For instance, if a change in the occupation density is suggested then the software will alter the number of toilets, cubicles, plumbing, disposal installations, points for power and IT together with the appropriate revised heating and ventilation installations. The program might even suggest that another lift be installed. Changes are automatically presented as text-based 'change reports' and can also be presented as colour graphs and diagrams. Live Options can be considered to be a leading software tool because of its ability to integrate both architectural and engineering services design cost drivers at the earliest stages of a project's design process. As a result, tools such as Live Options fill the crucial gap between rough cost estimates and more reliable project price forecasts based on full production drawings. As the project design progresses, actual data can replace assumed data and particular items such as those that are eligible for tax deduction can be tagged for particular attention. None of the developments described above replaces the professional cost consultant's skill and professional judgement that remains at the core of the building project price forecasting advice process (see Fig. 7.1). Issues such as the proposed process and method of procurement, the buildability of the scheme, and the sequence and allocation of risk, all need to be considered by expert building project price forecasters. The facility to interact with a modelling system such as Live Options allows the forecaster to assess the cost impact of issues such as anticipated project time-scales, indices, and individual rates, particularly for key items of a project's emergent design such as, say, steelwork arrangement or curtain walling specification. This interaction ensures that reliable good-quality early-stage building project price advice is capable of being determined.

In practice, it has been found that all project stakeholders have praised the building project price forecast reports generated by models such as Live Options for their clarity and the transparency with which all assumptions within the forecast have been made. This approach results in design teams and client organisations having the capability of interrogating and challenging key inputs in order to test out differing design scenarios prior to a client finally committing further resources to a project's design.

Live Options has been used on a number of national and international projects, one of the most demanding being for a new bank project in Uzbekistan. This proposed building was 26 storeys high and included three levels of basement. The shape of the building changed in area and plan at a number of different levels. The project report sheets had to be converted from their UK currency into US dollars using the indices within the system. Given the location of the project it was then necessary to consider the nature of the construction processes adopted within the local industry. As the local industry was labour-intensive in its approach, it was necessary to adjust this cost driver within the system to allow for the differing labour rates and productivity levels likely to be achieved on the site. The model then had to adjust for local taxes on imported building materials. Such adjustments were made via alterations to specific rates within certain of the elements included in the forecast. The finalised project report for the scheme was produced within a week and was well-received by the client. Such results illustrate that forecasts formulated with the aid of such explicit and transparent models aid the common ownership of the building project price forecast. This in turn helps ensure that the advice provided is seen as being reliable and as a result adds value to the building project's design processes.

7.6   The way forward

As society increasingly becomes aware of the significance of sustainability to the well-being of our future generations, so knowledge-based information management systems such as Live Options will have to be refined to an even greater extent. Such refinements will have to ensure that affordability of a particular course of action for a particular project can be assessed on more than just an initial first-cost basis.

Software tools such as Live Options bring together cost and design expertise in the crucial early stages of a project where often 80% of the design can be fixed in some 20% of the time available. Professional cost consultants with access to such tools will be in a position to continue to add value to the processes by which a client's requirements are converted into actual building project proposals.

As CAD systems improve, so the potential for systems such as Live Options to become more fully integrated into the design process increases. Such an integrated approach to 'expert system' development would enable practitioners to properly deal with the assessment of the cost impact of conceptual design and engineering designs in a resource-efficient manner. Full use needs to be made by cost consultants in general of the ever-increasing technological advances in both CAD software packages and knowledge-based information management systems such as Live Options. Also there needs to be consistent standards of measurement determined and a consensus formed on the optimum structure of the ensuing cost database between the major UK and overseas stakeholders connected with the development of such integrated systems. Such developments need to be backed by influential bodies such as the Confederation of Construction Clients, the RICS and RIBA in order to ensure implementation.

References

Bowen, P.A. (1995) A communication-based analysis of the theory of price planning and price control. *Research Paper Series*, **1**(2) Royal Institution of Chartered Surveyors.

Bowen, P.A. & Edwards, P.J. (1998) Building cost planning and cost information management in South Africa. *Journal of Construction Procurement*, June, 16–25.

Egan, J. (1998) *Rethinking Construction* (the Egan Report). Department of the Environment, Transport and the Regions, HMSO.

Fortune, C.J. & Hinks, A.J. (1998) Building project price forecasting models in use – paradigm postponed. *Journal of Construction and Property Financial Management*, March, 5–25.

Fortune, C.J. & Lees, M.A. (1996) The relative performance of new and traditional cost models in strategic advice for clients. *RICS Research Paper Series*, March, **2**(2).

Hinks, A.J. (1998) Conceptual models for the interrelationship between information technology and facilities management process capability. *Facilities*, **16**(9/10), 233–45.

8 Life-cycle/Whole-life Costing

Christine Pasquire and Lisa Swaffield

8.1 Introduction

This chapter will outline the principles and traditional techniques of life-cycle/whole-life costing, and discuss the relevance to modern procurement methods such as the UK private finance initiative (PFI) or build, own, operate (BOO) type projects, where financial risks have to be assessed over a far longer contract period. Some of the barriers to the successful implementation of the techniques will be discussed, along with some new research that is developing the traditional life-cycle costing techniques into more appropriate methods for adding value to modern construction projects.

The life-cycle cost (LCC) is defined as the present value of the total cost of an asset over its operating life, including initial capital cost, occupation costs, operating costs and the cost or benefit of the eventual disposal of the asset at the end of its life (RICS, 1999).

The term 'whole-life costing' is more recent. The most commonly adopted definition for whole-life costing appears to be (Robinson & Kosky, 2000): 'The systematic consideration of all relevant costs and revenues associated with the acquisition and ownership of an asset.'

The relevant costs include initial construction cost, fees, and interest, along with occupancy costs expected to be incurred in the future, such as rent, rates, cleaning, maintenance, repair, energy and utilities. The total cost for each possible option related to the project is calculated, with future costs discounted to present-day values. This varies little from the concept of life-cycle costing. The inclusion of revenues in the definition is new and points strongly to the need to identify the earning capacity or revenue stream for the building, introducing the idea that a building may be an asset rather than a liability. This is an extremely important concept to all organisations (whether commercial or otherwise) and requires a change in traditional corporate accounting methods if real benefit from such asset creation is to be felt.

Whilst asset creation is the overall goal for construction and property management, whole-life costing (WLC) is an important part in its achievement. A major driver in this shift of emphasis is the growth of public/private partnerships (PPPs) for the provision of public facilities and infrastructure, principally through the UK's PFI and similar BOO-type projects, which demand the consideration of revenue and expenditure over a relatively long term (20–25 years). Under this type of arrangement, the building and its smooth operation become a core business split from the process undertaken inside, as the two functions reside with separate organisations often associated with a transfer of ownership on completion of the specified term.

Clift, Lyncehaun and Cox (1999) take a slightly differing view, making the following differentiation between the terms whole-life costing and life-cycle costing.

Whole-life costing	Predicts the costs of use over the whole life of the building
Life-cycle costing	Estimates the cost over part of the building life, usually a single occupancy

It is, of course, very difficult to make decisions at the feasibility stage relating to the whole life of a proposed building with any degree of confidence, especially if it could change owner, occupier and function several times over a life of perhaps more than 100 years. There may be a significant difference in the potential life of the building structure and the intended life of a specific use. Whilst the period of a building's specific use by the original client may seem easier to deal with, it will not promote the flexibility of design and sustainability now desired on environmental grounds.

Any technique that attempts to account for future circumstances is inherently risky, as the future is uncertain. Logic dictates the longer the period of time accounted for, the greater the risks involved, and therefore the less accurate calculations are likely to be. It is important therefore to clearly identify the time span being accounted for. This may vary greatly; for example, a fast-food chain may have a three-year investment cycle whereas a cathedral would be expected to have a life of 100 years plus (Hoar & Norman, 1992). According to Ferry and Flanagan (1991), the life of a building and/or component can be expressed in functional, physical, technological, economic and/or social and legal terms. In the case of PFI-type projects, consideration of building life is no longer an option: private firms are expected to assess and price risks associated with long-term operation and maintenance of traditionally public sector buildings such as schools, hospitals and prisons. In such cases, it is important to recognise the difference between the whole life of the building and the intended occupancy or term of the contract.

The Building Research Establishment (Clift & Bourke, 1999) recommends WLC should be carried out at various stages of the procurement process:

- The initial investment appraisal or decision whether or not to build
- The assessment of feasibility of alternative construction solutions
- Outline design
- Detailed design (including choice of components and services)
- Tender appraisal
- Assessment of variations during the course of construction
- Final account
- Assessment of the effectiveness of the construction (post-occupancy evaluation)

The data created should be carried forward for use during the operational life of the building, contributing to the facilities management (FM) processes where cost and performance can be monitored. As changes occur in the building life and repairs/replacements become necessary, further WLC exercises will be needed. The optimum use of WLC therefore extends beyond the design and construction of the building into the operation and rehabilitation/disposal (Ferry & Flanagan, 1991).

8.2 Key concepts

The essence of life-cycle/whole-life costing is accounting for all possible costs associated with constructing and operating a building, and considering these costs at their present-day values. Therefore, the essential elements to understand include the following.

- *The time value of money/discounting (which incorporates allowances for interest and may include for inflation rate).* A simple explanation of this concept would be that a sum of money has a different spending power depending on when it is received, and money received today can be invested for growth so that it would have greater spending power in the future. For example, £100 received today could buy a Vehicle Excise Duty car tax disc (for a vehicle with engine capacity of 1.1 litres or less). To calculate how much money would be required now to have the same spending power in one year's time, it is necessary to consider the interest and inflation rates. £95 invested now at 5.5% would grow to £100.23 in a year. However, inflation at 3% would mean that £103 would be required to purchase the same item next year. This simple example does not take account of any other price changes that may occur in the intervening period, such as the Chancellor of the Exchequer increasing the tax to say £150 in the Budget. This illustrates one of the main problems with the technique of life-cycle costing: however carefully the calculations are performed, there is no way of accounting for any unforeseen changes that may occur in the future.

- *The building life* – either the intended term of occupancy by the original client, the contract period for which the private sector consortium is to assume the financial risk of the project in PFI or other public/private partnership arrangements, or the useful life of the building (irrespective of tenancies) for property investors. Building life clearly needs careful definition from the outset.
- *The costs expected to be incurred initially and during the building life, for the various design alternatives* – include energy costs, repair, maintenance and even such things as cleaning. It is important to define these costs clearly for PFI-type projects as liability for them will be split between the building manager (PFI consortium) and the building user (client) differentiated as the so-called hard and soft facilities management costs – hard being the upkeep of the building and its systems and soft being the 'hotel services' which would include daily, internal cleaning for example.
- *The design lives of the various components and equipment*, so that the calculations can include for replacement costs in the appropriate periods.
- *Any tax implications*, such as capital allowances available for developments in certain areas, or items of plant and equipment that qualify for capital allowances even when the building itself does not. These allowances are claimed against taxable profits, so the implications for non-profit-making organisations must be considered.
- *Deterioration* – where the building and its component parts age, becoming less fit for their purpose and slip into obsolescence.
- *Obsolescence* – where changes in technology, land values, society, legislation, fashion, etc. render the building unfit for its purpose during the intended design life, or make the economic life of the building shorter than the planned design life. Unforeseen obsolescence has a serious impact on the use of a building.

8.3   Techniques available for LCC/WLC

There are various methods available for performing calculations to compare options from simple payback (time required to return the investment) to more complex discounting techniques which take into account the time value of money. It is common practice to refer to published tables to find values within a wide range of interest rates for each technique but the formulae (Ferry & Flannagan, 1991) are given below for interest. Some of the techniques are summarised in the following subsections.

Simple payback method

This calculates the time required to return the initial investment, for example through income generated or perhaps energy cost saved. The option that has the shortest payback period is preferred. The method relies on simplistic assumptions about the preferred payback period and cannot adequately account for replacement costs to be incurred in different periods, resulting in rather ambiguous decisions as to the most appropriate solution. It also ignores any cash flow outside the selected payback period.

Discounting methods

Discounted payback method

The discounted payback method attempts to overcome a major disadvantage of the simple payback method, by considering the time value of money. Options with large net cash inflows in earlier years would be preferred over options where the same amount of cash would be received later. Likewise, an option that required £20 000 to be spent in year 8, would be preferred over one that required an expenditure of £20 000 in year 3, provided that all other things were equal.

Present cost

Present cost is the sum of money to be set aside in order to meet all eventual costs after allowing for the accumulation of interest. It is used where there are no tangible benefits to be taken into account and the formulae are:

$$PC_i = C_0^i + \frac{C_1^i}{1+r} + \frac{C_2^i}{(1+r)^2} + \ldots + \frac{C_t^i}{(1+r)^t} + \ldots + \frac{C_T^i}{(1+r)^T}$$

$$PC_i = \sum_{t=0}^{T=N} \frac{C_t^i}{(1+r)^t}$$

where C_t^i is the estimated cost for option i in year t, r is the discount rate, which is basically the difference between the rate of interest and inflation, and T is the period of analysis in years.

For example, a boiler forecast to require regular maintenance costing £200 per annum at today's prices with an assumed interest rate of 5% requires an investment today of £190.48 ($PC = 200/1 + 0.05$) for year 1, £181.41 ($PC = 200/1.05 \times 1.05$) for year 2 and so on. The reducing figure represents the present cost of £200 expended in the given year.

Net present value

Net present value takes into account both cost and benefit and is widely used commercially to appraise investment options. Large positive values mean profitable investments but for LCC costs may be represented without benefit, in which case the lowest negative value would be preferred. This would be an appropriate technique for use for cash flow comparisons for long-term projects such as PFI, BOO, etc. The formula is:

$$NPV_i = \sum_{t=0}^{T=N} \frac{(B_t - C_t^i)}{(1+r)^t}$$

where B is benefit in year t.

Internal rate of return

Internal rate of return (IRR) is the discount rate that gives a net present value of zero. This method is popular for development appraisal but has disadvantages for life-cycle costing calculations as the time-scales involved produce negative cash flow (cash outflow) and calculations are undertaken in the negative. For example, when comparing boiler A with an initial cost of £1000 and an annual running cost of £600 with boiler B which has an initial cost of £1200 and an annual running cost of £500, the calculated IRR would be 23.7%. Choosing boiler B over A would be prudent providing the required rate of return was less than 23.7%. The formula for calculating the IRR is:

$$IRR = i : \sum_{t=0}^{T} \frac{R_t}{(1+i)^t} - \sum_{t=0}^{T} \frac{C_t}{(1+i)^t} = 0$$

Benefit–cost ratio method

The benefit–cost ratio (BCR) method is similar to the IRR method above, but is suitable for cases where net outflows of cash are involved. The benefit–cost ratio is the ratio of the present value of future benefits to the present value of future costs.

The formula for BCR is:

$$BCR = \frac{NPV + I}{I}$$

where I is the present value of the investment costs of the project.

It can be seen there is a relationship between BCR and NPV and this is illustrated in Table 8.1, comparing two components with differing initial costs and annual outlays:

Table 8.1 Example of benefit–cost ratio.

	Type A	Type B	Discount factor 5%	Present value A	B	A – B	Present value R_t	C_t
Initial cost	(6000)	(7000)	1	(6000)	(7000)	(1000)	–	1000
Year 1 cash outflow	(1000)	(600)	0.952	(952.0)	(571.2)	400	380.8	–
Year 2	(1100)	(600)	0.907	(997.7)	(544.2)	500	453.5	–
Year 3	(1300)	(700)	0.864	(1123.2)	(604.8)	600	518.4	–
NPV				(9072.9)	(8720.2)			
Net benefits discounted							1352.7	
Net costs discounted								1000
NPV (B − A)					352.7			

$$\text{Benefit} - \text{cost ratio (B} - \text{A)} = \frac{1352.7}{1000} = 1.353$$

Annual equivalent value

Annual equivalent value can be used to compare options with differing life spans or where the life spans are not an exact multiple of the period of analysis. For example, a component requiring replacement every ten years would still have a five-year life on expiry of a 25-year occupancy/operating term. One method for taking this into account is by attributing a residual value to cover the remaining years of useful life. A less arbitrary method would be the use of an annual equivalent value. If two components have replacement cycles of n_A and n_B respectively, the first step would be to calculate the NPV for the individual life. Obviously these NPVs can not be compared directly and need to be expressed in equivalent terms. This is achieved by reading the present value of each life duration (n_A say five years, n_B say eight years) from annuity tables at a given discount rate (say 5%) to establish an annuity value, in this case 4.3295 and 6.4632 respectively.

The NPVs can now be expressed as annual equivalents by dividing them by the appropriate annuity value. For example, NPV_A

= £5200 and NPV_B = £6100 at five years and eight years respectively and an annuity rate of 5% gives an annual equivalent to component A of £1201.06 and component B of £943.80. Component B should therefore be chosen.

This is an important consideration for PFI-type arrangements where a residual value is to be paid when a facility transfers ownership.

Ranking and weighting techniques

These involve a structured approach to optimising the selection of components by considering a range of pertinent factors or attributes. The techniques revolve around the use of a comprehensive scoring matrix to compare options and allow the introduction of qualitative aspects into the comparisons. Examples of the techniques include:

- Weighted evaluation technique (Dell'isola & Kirk, 1981) enables overall design objectives to be listed and ranked to give a raw score. These raw scores are weighted and applied to a second phase which considers individual components or design solutions.
- SMART (Green & Popper, 1990) – simulated multi-attribute rating technique.
- QFD (Logothetis, 1992) – quality function deployment – developed for British Telecommunications as a tool for evaluating technical services. This tool (known as the House of Quality) can be readily adapted to evaluate and compare a wide variety of attributes and would be suitable for component comparison for WL/LCC.

8.4 Barriers to successful implementation of LCC/WLC techniques

Although the techniques for LCC/WLC have existed for many years, there has been relatively little success in applying these techniques to construction projects.

Availability of suitable data

The main point that can be noted from the methods is that, even though they appear scientific, with quite complex mathematical formulae in some cases, the model is only as good as the data it is based on. The data required include:

- Operational life/durability of components
- Costs associated with both planned and reactive maintenance/ repair
- Energy consumption
- Total installed costs

The sources of this data include manufacturers, FM consultants, building maintenance companies, in-house property management departments, contractors, utility companies, etc. However, whether the data exist in a usable form is another matter. Published data are sparse and not generally prepared for LC/WLC exercises. The main sources were identified as:

- Royal Institution of Chartered Surveyors Building Maintenance Information (BCIS – Building Cost Information Service)
- HAPM (Housing Association Property Mutual) *Component Life Manual*
- PSA (Property Services Agency) Tables
- Construction Price Books (various)
- Maintenance Price Books (various)
- BRESCU (Building Research Establishment's centre for promoting energy efficiency in buildings)
- Building Research Establishment: Centre for Whole Life Performance

Caution should be exercised when using component life figures as the information provided by manufacturers is usually based on accelerated tests, which may not be reliable (Clift & Bourke, 1999). The information is at risk as some components may require replacement earlier because of poor workmanship during installation or poor design detailing on jointing arrangements.

Fragmentation of the repair and maintenance process, along with the commercial sensitivity of cost data are major barriers to the collection of appropriate information for WLC. For some mechanical and electrical (M&E) services items, particularly IT, technological obsolescence may well have occurred before any cost feedback process produces useful data. Therefore, firms may have inadequate data, or may spend too much time collecting data and not enough time analysing options or exploring client needs.

Project finance

A traditional feature of construction projects is that they are often financed through clients' capital budgets, but the maintenance and running costs during the occupation phase come out of a different budget, usually under the control of an entirely separate department. This means there is no single overview of the building life costs and the main emphasis is placed on minimising the initial capital cost with little regard for the operating costs. The traditional focus on low initial costs is not appropriate for PFI-type projects, which require operating costs to be considered at the outset. It is logical that the importance of life-cycle/whole-life costing will grow as the use of these methods increases.

Short-term interest of client

The emphasis on lowest initial cost is even more prevalent when the construction and sale of the building is the core business of the client/developer – they are remote from the users, and tend to have little concern for the future operation and maintenance costs. However, public opinion is starting to change in favour of conserving resources, so life-cycle/whole-life costing techniques must develop to enable these preferences to be properly considered when evaluating options during the design phases of any construction project.

The short-term attitude of the client also applies to the case of business results. The benefits of whole-life costing are not normally realised by the business within the first 12 months (Robinson & Kosky, 2000). Any business under pressure to show good results to shareholders may consequently decide to reduce capital expenditure and put up with the greater running costs likely to be incurred in the future.

Professional fees

One reason why the preoccupation with initial capital costs prevails could be that the professional fees of consultants such as architects and quantity surveyors are often calculated as a percentage of the construction cost. It is difficult, therefore, for consultants to promote alternative costing approaches without radical changes to accepted fee practices.

Taxation issues

Robinson and Kosky (2000) identify Government fiscal policy, particularly the complex system of capital allowances, as a major barrier to the successful use of LCC/WLC in the UK. They found that, for tax purposes, capital expenditure on property is treated differently to revenue expenditure. Property running costs such as maintenance, repairs and energy are set against trading profit and receive relief at 100% against tax at the full rate in the year of expenditure. Capital expenditure, such as qualifying machinery and plant can only be offset against taxable income using a 25% writing-down allowance, calculated on a reducing balance basis. There is also no incentive for clients to invest in energy-efficient buildings, as the allowances relate only to plant and machinery, not to structures designed to reduce the need for mechanical ventilation plant, for example. A system of this sort does not encourage a balanced allocation of funds between capital and revenue expenditure.

In the UK, capital allowances apply only to profitable private companies that pay corporation tax, and exclude property developers who sell completed developments to investors, as these buildings are counted as 'stock-in-trade'.

There are several ways in which the UK Government could reform the system to encourage greater emphasis on whole-life costs for building projects, but it must be remembered that governments are unlikely to be overly concerned with the accounting practices of profitable private companies. It is up to the construction industry professionals, clients and researchers to take the lead in developing and promoting new whole-life costing techniques, and ensuring that clients receive best value within the constraints of the taxation laws.

8.5   LCC/WLC for PFI projects

PFI projects generally

The recent renewed interest in life-cycle/whole-life costing can be partly attributed to the emergence of public/private partnership procurement methods such as PFI and prime contracting, where the private sector has a long-term interest in the risks and rewards of the projects.

The Private Finance Panel (1992) define three types of project:

(1)   Financially free-standing projects, which the private sector is willing to undertake on the basis that it will regain its costs by charging the end-users – for example, toll bridges.
(2)   Joint ventures, which are partnerships between public and private sectors where the private sector retains control but both sectors contribute to the project and share its rewards. Public sector contribution could include transfer of existing assets, and its benefit could be non-monetary, such as the availability of an improved service.
(3)   Services provided to the public sector by the private sector, which tend to be services that involve significant capital

expenditure, such as prison provision where HM Prison Service does not own an interest in the prison buildings, but pays for a complete service provided to the inmates.

As different combinations of business expertise are required for different projects, various arrangements of contract have developed, these include:

- Design, build, finance and operate (DBFO)
- Design, construct, manage and finance (DCMF)
- Build, own, operate (BOO)
- Design, build, operate and maintain (DBOM)
- Build, own, operate and transfer (BOOT)
- Build, operate and transfer (BOT)
- Build, lease and transfer (BLT)

Depending on the type of contract, the private sector may take responsibility for all the design, building, financing, and operation of the facility for the contract period (typically 20–35 years), but the public sector always retains ultimate public accountability for the services provided. The private sector will not receive any payments before the facility is constructed and fully commissioned, so initially finances the capital expenditure. The theory is that the private sector is best placed to manage the risks, and that its expertise can lead to a lower overall cost and a better quality of service.

The typical structure of a UK PFI project is shown in Fig. 8.1.

PFI procurement stages and LCC

Table 8.2 summarises the key stages in PFI procurement and shows where life-cycle costing assessments will be required. Life-cycle costing should also be used throughout the PFI contract period to continually reassess the FM approach in terms of maintenance and renewal strategies of individual components, elements and viewing built assets as a whole; for example, assessing whether it is more economical to dispose of a poorly performing built asset and construct a new one. LCC therefore provides a mechanism to ensure that the most efficient method of service delivery is sustained for the duration of the PFI contract.

Value for money – in terms of price, quality, risk transfer and ancillary incomes (such as the sale of land or existing assets) – is a primary-aim PFI procurement. A public sector comparator (PSC) is used during the bid period to compare the value for money achieved through PFI procurement of services to traditional procurement and operation of fixed assets. A key feature of the PSC

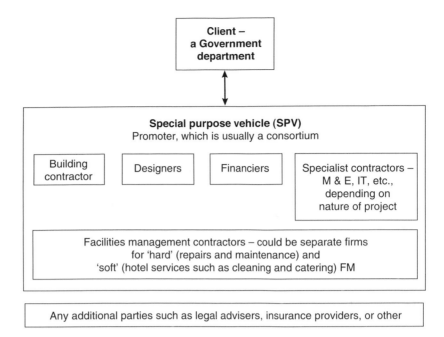

Fig. 8.1   Typical structure of a PFI project.

is that allowances are made for the value of additional risk transfer and also the higher cost of finance in the PFI method. The PSC must represent the costs of achieving equivalent outputs to those required by the project agreement/PFI contract.

Risk transfer in PFI projects

Risk transfer is crucial to the PFI deal. The public sector goal is for optimum risk transfer, which means allocating risk to the party best placed to manage it at least life-cycle cost, so that the key value for money requirement is not sacrificed. The private sector must genuinely assume risk, and clearly define the risk allocations in the PFI bid. Not all risks can be transferred to the private sector.

The risks involved in a PFI contract can be summarised as follows (Private Finance Panel, 1996):

- Design and construction risk
- Commissioning and operating risks
- Demand for volume/usage risks
- Residual value risks
- Technology/obsolescence risk
- Regulation and legislative risks

Table 8.2   PFI procurement stages and life-cycle costing exercises.

PFI procurement stage	Relevance of LCC
1. Establish business need	
2. Appraise the options	
3. Outline business case	
4. Creating the project team	
5. Publication of the OJEC notice	
6. Deciding tactics	
7. Pre-qualification of bidders	
8. Short-listing	Short-listing is a test of project-specific performance. It may require the private sector to provide details of which risks they are prepared to take and outline pricing, which would probably require a 'big picture' life-cycle cost estimate.
9. Revisit and refine the original proposal	
10. Invitation to negotiate (ITN)	The ITN would detail: • the service required, in output terms • the constraints on the scope of the project • proposed contractual terms (length, payment method) • the evaluation criteria for bids • the scope for variant bids A formal bid will be returned to the public sector client up to four months from the receipt of the ITN. Obviously a detailed life-cycle cost estimate of the costs associated with the construction and hard FM will have to be prepared by the private sector in order to calculate the unitary price.
11. Negotiation with bidders	The public sector will seek conformation of the risks transferred and subsequently further development of the detailed contract terms. The public sector client will also require assurance that the contract service outputs will be met. This stage is completed when the bidders issue their 'best and final offer'.
12. Selection of preferred bidder and negotiation to financial close	'The PFI proposition should be tested against key risk transfer, value for money and affordability criteria' (Private Finance Treasury Taskforce, 1999). The second-place bid offer will be kept at the table, presumably to maintain competitive pressure on the preferred bidder.
13. Award contract	Contract parties sign the project agreement and financial close is reached (the point at which the unitary charge is fixed).
14. Contract management	

Source: Clayton (2000).

A survey (Clayton, 2000) of PFI estimators from contractors and SPV organisations looked at the issue of risk transfer and some of the results are summarised in Table 8.3 below. It can be seen that there is a switch in responsibility between the construction contractor (who has only a short-term interest) and the FM contractor. This has an impact on the LCC/WLC process and may lead to the construction contractor being pressured to design for least capital cost in the traditional way despite all that has previously been said about PFI and the need to consider LCC/WLC. As Table 8.3 shows, risks associated with maintenance cost and major repair and renewal varied, with the following scenarios found:

- The hard FM contractor assumes the risk for all the costs associated directly with the 'upkeep' of the built assets.
- The costs of routine or low-value maintenance are the responsibility of the hard FM contractor, but the SPV (special purpose vehicle) takes the risk for the costs of major repairs and renewals (through a separate fund), usually employing the hard FM contractor to conduct the works.

Table 8.3 Parties responsible for various risks in PFI projects.

Risk	Party responsible for risk				
	Public sector client	SPV	Design-and-build contractor	Hard FM contractor	Other consortium member
Construction phase:					
Planning		✓	✓		
Design			✓		
Construction costs			✓		
Delays			✓		
Legislative/regulatory			✓		
Operation phase:					
Maintenance cost		✓		✓	✓
Major repair and renewal		✓		✓	✓
Penalties	✓				
Demand	✓				
Third-party revenues	✓	✓			
Residual value – built asset ownership	✓				
Residual value – built asset condition		✓		✓	✓
Technology and obsolescence		✓		✓	✓
Energy	✓	✓		✓	✓
Regulation/legislative – not service-specific		✓		✓	✓
Regulation/legislative – service-specific	✓				

- The costs of routine or low-value maintenance are the responsibility of the hard FM contractor, but the SPV and the hard FM contractor share the risk on the costs of major repairs and renewals (through a separate fund) with the hard FM contractor conducting the works.
- The hard FM contractor takes responsibility for routine maintenance and a separate consortium member takes responsibility for the cost of major repair and renewals.

Problems may occur with defining which work could be described as 'routine or low-value maintenance' and which was 'repair and replacement', and gaining access to the relevant funds.

Discounting

The principles of discounting and the time value of money were discussed above. Selecting the appropriate discount rate is vital to the validity of life-cycle costing calculations. In PFI projects, the discount rate is generally determined by the client in the first instance, to retain parity of bids until one preferred bidder has been selected. When the preferred bidder has been selected, the SPV would then negotiate a discount rate with the public client before the final unitary price was agreed, and inform the building contractor's LCC estimators which discount rate to use. LCC exercises completed at each project phase would generally be submitted to the SPV without discounting, for inclusion in the overall financial model.

Indexing the unitary charge

The unitary charge that the client pays the SPV for the facility is usually index-linked to the retail price index. Therefore, the private sector is left with the risk of differential inflation between the index used and the actual project costs. This is important to the PFI consortium, as the WLC calculation must balance expenditure against revenue. As a commercial entity, the PFI consortium will wish to maximise profit. They will usually be required to provide the finance for the construction cost before the revenue stream is established and it may be preferable for this first cost to be minimised rather than optimised. WLC exercises can be used to do this along with optimising the later, and possibly higher, expenditure needed as a result of the lowest first cost approach. However the PFI consortium plan to manage the finances, WLC should have an important role to play for project appraisal, initial tendering, detailed tendering, design development, subcontract tender appraisal, estimating finance requirements, development of maintenance strategy, liability-to-date calculations, budget forecasting, and maintenance and renewal strategy appraisal.

Running and maintenance costs in PFI contracts

Mechanical, electrical and plumbed services have significant running and maintenance costs in most buildings, whether PFI contracts or not. However, in PFI contracts there is the additional implication of incurring a service penalty charge if an operation failure occurred in the M&E services, as it would affect the quality of the PFI facility provided to the public sector client.

Internal fabric items could have higher life-cycle costs on PFI projects than on traditional buildings, because the types of buildings that are provided under PFI contracts, such as army barracks, schools and prisons, are likely to have higher incidences of internal damage – either by accident or deliberate vandalism.

Application of LCC/WLC for PFI projects

The detail and format of an LCC/WLC study will vary greatly according to who requires the information, for what purpose and at what stage during the building life cycle. There is considerable demand for a single sum annual cost for early stage decision-making and public sector clients such as health authorities/trusts have collected cost data for preparing such a figure. This information is used as a base against which to assess new build options and is effectively the benchmark the PFI consortia are tendering to. Examples can be found on the Treasury Task Force PFI library web pages at http://www.hm-treasury.gov.uk

The tendering PFI consortia will undertake LCC/WLC studies and it is important these studies are coordinated and integrate properly as they will derive from the various contractors involved in preparing the construction costs and will need to be passed on to the FM companies taking up the operational risks.

The completed LCC/WLC study will only be as good as the initial briefing dictates and clear objectives relating to the spread of expenditure must be given to the contractors so the overall running and maintenance cost cash flow is properly balanced against revenue. This will enable appropriate measures to be taken such as the establishment of sinking funds, etc. Without this clear brief, PFI projects are in danger of reverting to the lowest first cost approach prevalent in traditional construction procurement.

The individual contractors involved will undertake their LCC/WLC studies on a component basis, usually dealing with the significant cost items only. There is currently no universal classification system for LCC/WLC data and research is urgently needed in order to develop a standard approach and implement best practice. In practice it appears that the structure and fabric is likely to follow an elemental classification, e.g. BCIS, SfBS, whilst building services may use a system basis, e.g. lighting, air-conditioning, security, lifts, etc. At present, any studies undertaken use ad hoc information classifications influenced more by the existing format of data than any informed and organised system.

Flanagan *et al.* (1989) proposed a hierarchical structure to LCC data that could be adapted for use with the LCC process defined by Kirk and Dell'Isola (1995). The information used would typically include that shown in Fig. 8.2.

GENERATION OF
ALTERNATIVES

- Siting and orientation
- Special features
- Configuration
- Subsystems
- Equipment
- Materials
- Methods
- Constructability
- Time
- Functional use
- Energy optimisation
- Marketing
- Adaptability

SITE DATA
- Climate
- Environmental

FACILITY COMPONENTS
- Initial cost
- Useful life
- Maintenance
- Operational

SPECIFIC PROJECT INFORMATION
- Programme
- Criteria and standards
- Operational mode
- Quantities
- Economics
- Miscellaneous data

Fig. 8.2   Heirarchy of LCC data.

Consideration of these factors leads to a variety of life-cycle cost predictions including:

- Initial
- Operations (energy)
- Maintenance
- Alteration/replacement
- Associated tax elements
- Salvage value

In addition to the above considerations there are a number of non-economic comparisons in balancing the component costs with the functional, technological and aesthetic factors of the project.

Individual studies are lengthy and detailed and extensive case studies of typical projects have been published, e.g. Hoar and Norman (1992), along with proposals for work sheets and analytical techniques, e.g. Kirk and Dell'Isola (1995) (see Further reading section for more examples). None of these refer specifically to PFI-type projects but the principles, techniques and processes apply equally to all projects regardless of procurement route.

An LCC exercise would typically include the following seven steps (Ferry & Flanagan, 1991):

(1) Set objectives, constraints and alternatives
(2) List assumptions and procedures
(3) Collect data
(4) Choose techniques, compute and discount cash flows
(5) Compare results
(6) Evaluate for uncertainty and risk
(7) Report findings and recommendations

The important aspect as far as PFI projects are concerned is the transfer of the initial LCC studies into a working facilities management tool to support both planning and costing the operation of the facility for the PFI consortium. This would form an eighth step in the above procedure. However, there is currently no guidance on how this should be achieved and this does not as yet form a focus for research in the public forum (individual companies view this work as commercially sensitive and are developing systems and procedures in confidence).

8.6   Research and the future

It has been shown above that there are several problems with current LCC/WLC techniques and their successful application to the construction and property industries, including:

- Availability of suitable data
- Project finance
- Short-term interest of the client
- Professional fees
- Taxation issues

Construction is a dynamic industry, where improvements are constantly being made in construction methods, materials, M&E and IT systems, business processes, procurement methods and management techniques. Therefore, it is only natural that costing methods, including LCC/WLC methods, should develop to keep up with other changes.

The culture of the construction industry has started to change towards greater trust and teamwork between the parties involved, with more collaboration and cooperation. There have also been technological advances such as the internet, email, and electronic data exchange generally. The combination of these changes means that there has been no better time than now to introduce new LCC/WLC methods, or a shared national whole-life cost database accessible on the internet. Anything is possible, but obviously detailed research is required into any new techniques to ensure the information collected is useful. The barriers to advances remain commercial secrecy over cost data and intellectual property rights over research.

Fuzzy logic and artificial neural network research

Research is being conducted at the Robert Gordon University involving the development of an integrated LCC model based on artificial neural networks and fuzzy set theory. Fuzzy sets have a gradual transition from member to non-member, allowing the modelling of uncertainty associated with vagueness, imprecision and/or lack of information regarding some elements of a given problem (Kishk & Al-Hajj, 1999). Therefore, fuzzy sets and fuzzy logic are ideal for consideration of LCC/WLC models, where data may be vague or have missing elements.

Artificial neural networks are models that work in the same way as the human brain (although in a greatly simplified manner), where each neuron receives information from other neurons, processes it and passes on an output to other neurons. Specialist computer software packages are available that can perform the calculations, but each network must be 'trained' with appropriate data where the software 'learns' about relationships in the data, so it can derive values for missing elements in future data sets. A great strength of the application of artificial neural networks to cost research is the ability to deal with non-linearity, which obviously could not be done with earlier techniques such as multiple linear regression analysis.

Recent developments have involved models that combine the benefits of both fuzzy logic and neural networks – neurofuzzy systems, where fuzzy logic can be used in the pre- and/or post-processing of data, especially less tangible, subjective data; and an artificial neural network is used to extract the required rules and membership functions for a fuzzy system (Kishk & Al-Hajj, 1999).

The proposed Robert Gordon University integrated LCC model aims to reduce the subjective elements that are predominant in life-cycle costing, and make LCC estimates quicker and cheaper to produce (Kishk & Al-Hajj, 1999). It is easy for researchers to proclaim the benefits of their proposals, but all too often the best research ideas fail to become adopted industry practice. This is either because researchers cannot gain access to adequate data for analysis – or training networks in this case – or because industry professionals do not trust these new research techniques that they do not understand and prefer to retain the traditional methods, unless their clients specifically request something different.

It is time that consultants and clients begin to adopt the same trust and teamwork approach with academic researchers, as they are now starting to do with contractors. When all the major players are pulling in the same direction, the industry can move forward.

Partners in Innovation Research

The Partners in Innovation Research programme has recently funded a number of research projects that involve whole-life/life-cycle costing; these are listed in Table 8.4. The fact that such projects are being undertaken shows that there is a definite commitment from the construction industry to improve the techniques and use of whole-life costing. The project summaries in Table 8.4 show that new approaches are being taken, with clients, M&E engineers and facilities managers taking lead roles. Further information regarding completion dates, contact details and final reports can be found on the DETR web site at http://www.dti.gov.uk/construction/index.htm

Traditionally in the UK, anything to do with construction cost was the domain of the cost consultant, but it appears that there are now new parties getting involved in the LCC/WLC area – investing time in research, trying to develop better techniques, encouraging standard approaches and better data collection, and promoting the benefits of LCC/WLC to add value to the project. Perhaps this is an indication that traditional LCC/WLC techniques were underused because they were unsatisfactory, and that new research and development is long overdue to benefit modern construction clients.

8.7   Value for money

Value for money is the optimum combination of whole-life cost and quality (or fitness for purpose) to meet the user's requirement (Procurement Policy Unit, 1998). The value of the cost advice service offered to construction clients during the early design stages can be improved by increasing the quality of the service provided in relation to the fee. The use of WLC/LCC would be one way of

Table 8.4   Recent WLC/LCC research projects.

Title	*Promoting Innovation In Construction – The Role of Facilities Management*
Contractor	BRE collaborative
Objectives	1. To reduce building whole-life costs through a proper consideration of management issues during the design process.
	2. To develop a better understanding of the role of facilities management in reducing building life-cycle costs and the opportunities for FM input to the building design process.
	3. To evaluate and promote the benefits of considering building management issues at the design stage.
	4. To seek out and promote examples of innovation in design that have resulted from a consideration of management issues.
	5. To identify further opportunities for innovation.
	6. To develop and promote an innovation strategy for the FM profession.
Title	*Whole Life Costing Guidance for Clients and Designers*
Contractor	Construction Clients' Forum
Objectives	1. To give potential users confidence in the application of a whole-life costing approach by applying it to actual projects, evaluating benefits, identifying problems and proposing solutions, and by producing joint CCF/BRE guidance for clients and their design teams.
	2. To guide the development of a standard approach to WLC.
	3. To help establish a practical regime for the assessment of the service life of building components, and initiate the creation of the associated data sets.
	4. To provide initial entries in a database of WLC projects through which UK progress in reducing the level of building costs, as measured by WLC, could be monitored.
Title	*Whole Life Costing Techniques Applied to Building Environmental Services*
Contractor	W S Atkins
Objective	To produce a guidance note, based on a case study, to inform and encourage building developers and designers to adopt a whole-life costing approach to the design of building environmental services by demonstrating the value, advantages and potential barriers involved in such an approach.

adding to the quality of the cost advice service, ensuring that the best long-term solution is adopted, ultimately achieving better value for the client.

The client's budget for construction costs is sometimes negotiable, depending on the type of client and the occupancy intentions. Some clients would be prepared to pay slightly more for a higher specification of materials or equipment, if it could be demonstrated that maintenance, cleaning, repair and/or replacement costs would be reduced over the intended occupancy period of the building. Cost advisers must embrace new ideas and techniques where they will enable clients to consider the relative importance of various attributes, and identify the optimum solution.

References

Building Cost Information Service, Royal Institution of Chartered Surveyors, subscription service, 12 Great George Street, Parliament Square, London.

Clayton, S.I. (2000) An investigation into life cycle costing for private finance initiative projects. Undergraduate dissertation, Department of Civil and Building Engineering, Loughborough University.

Clift, M. & Bourke, K. (1999) *Study on whole life costing*. BRE report No. CR/366/98, Building Research Establishment.

Clift, M., Lyncehaun, R. & Cox, D. (1999) *Movement for Innovation – Report of Technical Day 5 CIRIA*.

Dell'Isola, A.J. & Kirk, S.J. (1981) *Life Cycle Costing for Construction Professionals*. McGraw-Hill.

Ferry, D.J.O. & Flanagan, R. (1991) *Life Cycle Costing – A Radical Approach*. Construction Industry Research and Information Association.

Flanagan, R., Norman, G., Meadows, J. & Robinson, G. (1989) *Life Cycle Costing: Theory and Practice*. BSP Professional Books.

Green, S.D. & Popper, P.A. (1990) *Value engineering: the search for unnecessary cost*. CIOB Occasional Paper 39.

Hoar, D. & Norman, G. (1992) Life cycle cost management. In: *Cost Consultancy Techniques: New Directions*. Blackwell Scientific Publications.

Kirk, S.J. & Dell'Isola, A. (1995) *Life Cycle Costing for Design Professionals*, 2nd edn. McGraw-Hill Inc.

Kishk, M. & Al-Hajj, A. (1999) An integrated framework for life cycle costing in buildings. Paper given at *RICS Research Foundation COBRA Conference*, The challenge of change: Construction and Building for the New Millenium, 1–2 September 1999. University of Salford, pp. 92–101.

Lotothetis, N. (1992) *Managing for Total Quality*. Prentice Hall.

Private Finance Panel (1996) *Private Finance in PFI Contracts – Practical Guidance on the Sharing of Risk and the Structuring of PFI Contracts*. HMSO.

Private Finance Treasury Taskforce (1999) *Taskforce Guidance – 14 Stages to PFI Procurement*. HMSO.

Procurement Policy Unit (1998) *Procurement Policy Guidelines*. HM Treasury, HMSO.

Property Services Agency, *Tables*. HMSO.

RICS (1999) *The Surveyors' Construction Handbook – Section 2: Life Cycle Costing*. Royal Institute of Chartered Surveyors.

Robinson, G.D. & Kosky, M. (2000) *Financial Barriers and Recommendations to the Successful Use of Whole Life Costing in Property and Construction*. Construction Research and Innovation Strategy Panel (CRISP).

9 Environmental Management

Jon Robinson

9.1 Introduction

In order to establish the subject matter of this chapter, a definition of environmental management is attempted. The environment comprises the 'surroundings in which an organisation operates, including air, water, land, natural resources, flora, fauna, humans, and their interrelations' (Standards Australia & Standards New Zealand, 1999). Management is the 'conduct and control of organised human activity' (Calvert, *et al.*, 1995). For the purposes of this chapter, the organisation is taken to be a parcel of urban property, i.e. land and buildings, together with the activities of its occupants. Thus environmental management is defined in the context of this chapter as the conduct and control of urban property within the environment. Environmental management includes the impact of the property on the environment and vice versa and thus it subsumes ecology. It is the activity that makes possible the concept of sustainable development, a concept that 'stresses using resources of energy and materials in a responsible way so that future generations can benefit from them too' (Fryer, 1997).

The process of environmental management is continuous throughout the life cycle of the property. The life cycle comprises 'consecutive and interlinked stages of a product system, from raw material acquisition or generation of natural resources to the final disposal' (Standards Australia & Standards New Zealand, 1998). In the context of property and construction, it is a time horizon, which commences with the acquisition of land, includes the design and construction of buildings, continues with the ongoing operation of the property, and ceases with the ultimate demolition or deconstruction and recycling of the property.

Environmental management in property and construction covers a wide range of activities. Broadly, these may be arranged under several classifications; for example (Cooper, 1996):

- The upstream production of the resources and materials used in a project

- The operation of the project through its life cycle
- The downstream use and/or disposal of the project at the end of its life cycle

A second classification comprises the stages throughout the project life cycle from inception through to disposal:

- Inception and design
- Delivery and commissioning
- Operation
- Disposal, including demolition, recycling, redevelopment, renewal and upgrading

Life-cycle costing methods have been used in decisions about property and construction options for many years (Stone, 1980; Flanagan & Norman, 1984; see also Pasquire & Swaffield chapter 8 in this book). Levels of complexity of life-cycle costing have been recommended: the experiential, the feasible and the technical (Robinson, 1996). Experiential life-cycle costing amounts to a statement of good practice wherein, for example, unsustainable elements are deleted from the designer's repertoire. Feasible life-cycle costing combines cost and value analysis procedures with an economic feasibility study. The more detailed technical approach is illustrated elsewhere (Robinson, 1996) in which the costs and benefits associated with a series of options is analysed including the effects of taxation, externalities and discounting and inflation.

A third classification consists of the environmental issues which, for example, have been grouped under the following headings by the Construction Industry Research and Information Association (Fryer, 1997):

- Energy use, global warming and climate change
- Resources, waste and recycling
- Pollution and hazardous substances
- Internal environment of buildings
- Planning, land use and conservation
- Legislation and policy issues

The purpose of the first part of this chapter is to provide a discussion of these issues, leading to the types of management decisions that need to be made at stages in the life cycle of a property in order to minimise its impact on the environment. The optimum basis for sustainable environmental management is in the provision of a facility of high environmental design. The second part of the chapter is a brief case study concentrating on the decisions made at the inception stage, which will affect management outcomes throughout the life cycle. Environmental impact assessment techniques and environmental management systems (Griffith, 1994; Standards Australia & Standards New Zealand, 1996) are considered to be beyond the scope of the chapter.

9.2 Issues

There are at present two schools of thought about the effect of greenhouse gases on global warming and climate change. The first is that greenhouse gases are changing the climate and the second is that there is no evidence that greenhouse gases are having any effect on climate (Scott, 2000). However, this chapter adopts the conventional wisdom as enshrined in the Kyoto Protocol, i.e. that greenhouse gas emissions must be reduced. This in turn requires *inter alia* a reduction in energy use. Energy use in buildings may be classified as embodied energy, operational energy and transport energy, all of which contribute to greenhouse gas emissions. Thus the carbon impact of a proposed building must also be modelled.

Embodied energy

Embodied energy is the energy associated with the production and transportation of the materials used in the building structure, services and finishes. A new building will have a relatively high embodied energy component whereas a recycled building will have a relatively low embodied energy component, as the energy associated with the existing structure and retained components is not counted.

The construction of a building, although a one-off event, involves an intensive use of resources, in particular materials and energy. Embodied energy is estimated to be equivalent to 10–20 years of operational energy (Aye *et al.*, 1999; Fay, 1999). Whilst for dwellings, operational energy exceeds capital energy in about the twentieth year of building life (Fay, 1999), it is estimated to be shorter for commercial buildings which are more subject to economic forces (Aye *et al.*, 1999). Capital energy includes building disposal and recycling, which combines the dismantling and reprocessing of materials, and detailed figures have been produced (Jacques, 1996).

Operating energy

Operating energy is that associated with the running of the building systems, such as heating and ventilation, air-conditioning, lighting and power and vertical transportation. It includes base energy loads such as security lighting and variable loads which the owner or tenant may seek to minimise. Electricity, because of its great versatility and relative safety, is the major form of operational energy. Analyses undertaken to demonstrate the potential savings in operational energy show that substantial savings may be made in both energy use and life-cycle cost terms when comparing a typical building design with an energy-efficient building design (in the order of 50–70%) (Aye *et al.*, 2000).

Operating energy assessments are made having regard to the electricity and gas requirements of the users in a suitable building. The assessments assume that the energy design targets set by organisations such as the British Property Federation and the Building Owners' and Managers' Association (now the Property Council of Australia) are met (BOMA, 1994, 1996; Aye *et al.*, 2000).

Transport energy

Transport energy is that energy associated with the location of the premises. If a location with good access to public transport is selected, then the transport energy will be lower than a location requiring a large proportion of the staff and visitors to arrive by car. An assessment of the environmental impact of the proposed project requires that transport energy be taken into account (Prior, 1993).

Tenants and visitors usually access commercial buildings five to seven days every week. In a city, the normal occupants of a commercial building choose a variety of means to commute to and from their workplace. During a workday they will also take excursions from the building for meetings or for the delivery of goods and services. Visitors will similarly attend the building in a variety of ways, mostly by carbon-based transport modes. Over a building lifetime, many passenger-kilometres are accrued, amounting to a significant energy use. The location of the building relative to the dwellings of its users and the nature of the commercial activities it supports leads to the quantification of the transport energy and consequent carbon use of the building (Evans, 1992).

Carbon

Whilst energy consumption has been of concern ever since the energy crisis of the early 1970s, the current global focus is now on the environmental impact of energy use. Fossil fuels supply a clear majority of our current energy needs. Carbon dioxide (CO_2) a major product of the use of fossil fuels, has been described as the major element in climate change through its accumulation in the earth's upper atmosphere leading to global warming. At present, the property and construction sector accounts for approximately one-quarter of all energy-related emissions in developed countries (Australian Greenhouse Office, 1991).

Most developed nations are now considering ways of attending to their Kyoto targets of reducing by 10% their 1990 levels of greenhouse gas emissions by 2010. One way of dealing with the target is to focus on carbon sinks as a means to absorb carbon entering the atmosphere. Trees are known to bind carbon. It is thus suggested that emissions targets can be addressed by the planting of trees (Kirschbaum, 1996). The amount of carbon absorbed and the effectiveness of tree plantations as a carbon sink depend upon the type of trees, density of plantation, location, age, climate and soil. This activity is attractive to a country like Australia which has plentiful land, excessive land clearing, a need to address erosion and salinity problems and a heavy reliance on fossil fuel production, use and export. It is not so attractive to smaller highly populated countries like Japan which is acquiring forest and plantation land elsewhere.

9.3   Resources, waste and recycling

Environmental design

A building of high environmental design is one in which the use of energy to modify the internal environment is kept to a minimum. This is accomplished by adopting passive solar design in which the building itself contributes to a lower requirement for energy. There are several ways in which passive design takes place. First, there is the potential to create thermal mass by the use of heavyweight masonry construction, either reinforced concrete or brickwork. The mass acts as an insulation material, which protects the internal space from the external climate. Second, there is the use of lightweight insulation material that may be placed in hollow walls or in roof spaces. Third, there is the orientation of the building to take advantage of insolation, which can act as a heat source if buildings are oriented so that substantial glazing faces south (in the northern hemisphere and north in the southern hemisphere). The use of associated sunshading devices is required in summer to protect the building against excessive heat gain but these should be arranged in such a way as to let sunlight through in winter.

Ecology

All construction activity involves some loss of ecosystems, either directly through site clearance for construction or indirectly through materials and component manufacture and production. Materials which are non-renewable or slow to replace should be avoided for maximum environmental results. For example, it is generally accepted that designers should avoid the use of rain forests because deforestation leads to loss of habitat as well as loss of the carbon sink. Furthermore, it is recommended that recycled materials be adopted where appropriate. Development features leading to land degradation by way of erosion or contamination should also be avoided. Where land has been disturbed for mining or other development, a policy of regeneration should form part of the environmental management during the life cycle of the project. It is important to protect the native flora and fauna that may be affected by a proposal.

Construction methods

Differing types of construction methods have different levels of environmental impact. A process of pre-qualification is required for best practice to establish that consultants and contractors competing to take part in the project have suitable experience in environmentally sustainable development. This pre-qualification process will require consultants to design and document a project having undertaken energy and carbon modelling and life-cycle cost and energy analysis. It will also require contractors to adopt sustainable construction methods and sustainable materials and put in place an on-site environmental management process. The principles of partnering and alliancing should be adopted as best practice in order to achieve optimum environmentally sustainable projects. Particular attention would be paid to the level of waste, which varies from trade to trade, and to the types of materials used, although many of these will be selected during the design phase. For example, if a potentially hazardous material such as medium-density fibreboard is specified, then the contractors will need to arrange a restricted access cutting room that is properly ventilated. In order to minimise the noise associated with construction, contractors may need to hire cranes and other plant driven by relatively quiet electric motors rather than noisy diesel engines. Environmental sustainability involves occupational health and safety not only during the construction process but also during the whole operational life cycle.

Demolition

The end of the life cycle needs to be considered from the outset. The disposal of the improvements to the property should be considered from the point of view of resource optimisation. If a building is to be completely demolished and the land redeveloped, the materials should be at least capable of being used as clean landfill. A better solution environmentally is to make the most of recyclable materials on site by retaining as much of the original structure as possible. This requires initial design for flexibility and adaptability but the requirements of a cyclical market tend to result in complete demolition of existing structures in a boom or the decommissioning of buildings in a recession. The retention of a building structure, which only partially reflects the requirements of a new occupier, may be counter-productive. This leads to a consideration of industrialised building in which demountability of the building elements could assist in the recycling process.

The property market requires that buildings be continually refurbished or redeveloped to maintain market position and financial value. This militates against environmental principles as materials and buildings are often demolished long before the end of their service lives. However, unless refurbishment takes place, a building may well become obsolete and vacant as occupiers move to newer buildings in better condition. The functional life of buildings in a market must not only take into account the physical or service lives of the components but also the economic life. Thus design for flexibility and adaptability becomes important. Not only should buildings be designed to take account of constructibility, the design should also take into account deconstructibility to enable easy demolition and recycling of buildings and building components.

9.4 Pollution and hazardous materials and substances

A number of materials used in building and building operations have the potential to cause significant health problems during construction and during operations throughout the life cycle. In addition, industrial pollution may be the outcome of the manufacture of materials used in construction. The building itself may be required for an activity that may be less than environmentally friendly, e.g. a chemical plant.

Hazardous materials

The major hazardous materials used in building are asbestos, formaldehyde, volatile organic compounds, lead and pesticides. Many gases result from cooking or heating or smoking in domestic situations and do not often occur in commercial buildings. Asbestos is no longer used in buildings but may be encountered in older buildings where refurbishment or maintenance is required. Most jurisdictions require that competent contractors are engaged, prior to any work, including demolition, being undertaken to remove asbestos. Formaldehyde is primarily found in reconstituted wood products such as medium-density fibreboard. As discussed above, this material is sawn in special spaces on building sites to avoid health problems. A group of chemicals referred to as volatile organic chemicals are present in a large number of synthetic products that are used in buildings, such as paints, adhesives, sealants, wood products and carpets (Brown, 1997); they are also found in cleaning products. Lead is found in older buildings, particularly in paint and this needs to be considered during the refurbishment of older buildings. Finally, some potentially hazardous pesticides and herbicides are used in and around buildings. All of these products can contribute to sick building syndrome (see below).

Contaminated land

Industrial land in urban areas can become contaminated with hydrocarbons, heavy metals, nuclear waste, organic waste, lead, mercury, arsenic and other toxic materials. Development proposals for brownfield sites (industrial land that has become unused) must take into account the likelihood of contamination and its remediation. It is arguable that urban consolidation should be promoted as a way of remediating industrial contamination of land.

9.5   Internal environment of buildings

Heating, ventilation and air-conditioning

Although a major thrust of environmental management is the promotion of passive solar systems, the property market requires some active control over the internal environment. Therefore, heating, ventilation and air-conditioning (HVAC) systems are likely to remain a major element in buildings, particularly non-residential buildings. The significant element here is in the comfort of the building occupants. Completely passive systems may render a building unusable on either very hot or very cold days. The nature and quality of installation of an HVAC system will affect air quality and energy consumption; for example, levels of carbon dioxide, carbon monoxide and tobacco smoke can become highly toxic in low-quality systems. A properly engineered variable air volume system is preferable in terms of energy conservation. On the other hand, an inappropriately arranged terminal reheat system requires a higher energy consumption.

Indoor air quality

The materials used in buildings are often the source of contamination as discussed above. However, the way in which buildings are used and maintained can also be a cause of contamination leading to building-associated illnesses which are catergorised into building-related illnesses and sick building syndrome. Sick building syndrome covers a range of conditions from allergies and minor irritations to low-level malaise including fatigue, headaches and nausea. Building-related illnesses are much more specific and include potentially fatal afflictions such as asthma and legionnaires' disease.

Two groups of contaminants have been identified (Williams, 2000). The first group comprises particles and biological contaminants such as dust, wetness and legionella. They are manifested in fresh air and return air and as a result of cleaning practices. All of these occur as a result of poor installation and/or maintenance and most of them can be eradicated by correct building maintenance procedures. The second group includes gas and vapour contaminants such as the volatile organic compounds and formaldehyde. As well as the environmental ethic, there is a legal liability for poorly maintained buildings through occupational health and safety legislation.

Research (Williams, 2000) has shown that many design inadequacies are present in air-conditioning systems in modern buildings. Particularly important is the location of fresh air intakes in relation to discharges from kitchens, toilets, boilers, car parks, cooling towers and dust and industrial pollutants from nearby sites. Also important is the nature of various activities and installations within the building such as cleaning materials and vapours from photocopiers. The quantity of fresh air circulating through internal spaces is also very important and can be affected by insufficient supply, location of air registers or overuse of return air.

Of paramount importance is the appropriate maintenance of the air-conditioning systems. Cleanliness and dryness are basic requirements and they are all too often ignored in the maintenance process. Accessibility to ducts and air handling units is often restricted by poor design and should be considered. Suitable filtration systems are required to ensure all bacterial and particulate agents are removed.

Another factor is the ability of building occupants to control the heating, ventilation and air-conditioning systems. Inappropriate location of thermostats or insufficient numbers of thermostats and zones can lead to a poor-quality internal environment. Most of these inadequacies arise from low first-cost requirements under the assumption that future costs are passed on to building occupiers.

Natural and artificial lighting

Most commercial buildings are designed to have a high level of artificial illumination, requiring a substantial energy input. This situation occurs with little regard for the availability of natural lighting which can be enhanced by thoughtful design, leading to a suitable building configuration. Furthermore, building occupants tend to have the lighting on whether or not it is required, typical levels being 500 lux at a hypothetical workface. It may be that an optimum is achieved with background lighting overall and individually operated (desk) lamps providing sufficient task lighting where it is required. Moreover, devices such as sensors that cause lighting to switch on and off as occupants enter a room or area, or timing devices that switch lights off automatically at certain times of the day, can help to reduce energy consumption as a result of lighting.

Hydraulic services

Environmental sustainability requires that the use of water in a building and the discharge of wastewater be reduced to an optimum level. A reduction in mains water use may be achieved by the installation of rainwater tanks and by the reuse of water on the site. This will also reduce the requirement for stormwater drainage. The on-site treatment and reuse of grey water and black water may achieve a reduction in wastes. The installation of landscaping in property developments provides a useful destination for the effluent.

In terms of water supply, modelling of the supply of and demand for water is required. Natural water supply is of course from rainwater and average annual figures for any location are readily available. It may be stored in tanks, but an allowance must be made for evaporation and these figures are also available. Demand modelling is undertaken on the basis of the average requirements of the projected building population. Water is required for human consumption and therefore a source of potable water is required. Water is also required for washing and the frequency of showering and hand washing needs to be assessed. Certain technological elements are available to reduce consumption, such as efficient showerheads and automated taps. In addition, water is required to operate toilet fittings into the waterborne sewerage system. Again, devices are available to minimise the use of water in such systems, including dual flush toilets and dry composting toilets. The treatment of wastewater is also possible using aerated treatment systems and filtration systems, but the effluent from these systems is also an issue and they generally require constant maintenance and cleaning.

Vertical transport

Vertical transport is a requirement both of the market and of regulations relating to access for the disabled. From a market point of view, the waiting time and the ability to lift an up-peak load is important. From a universal access point of view, all persons need to be able to access all levels of a building. An environmentally designed building will have these points of view optimised; for example, design layouts can place stairs in locations which encourage occupants to use them, with the lifts located in slightly less accessible locations, so the preference or ease of taking the stairs over lifts is subjectively reinforced.

Acoustics

The acoustic treatment of buildings is also an environmental requirement. This must minimise hearing damage to building occupants and will avoid distractions during working hours with a consequent effect on business efficiency. Materials specification and insulation may be used to prevent outside noise penetration and party-wall noise penetration. Materials specification may also be required to optimise internal reverberation to avoid the currently popular use of hard surfaces creating a continually high level of internal noise pollution.

9.6   Planning, land use and conservation, including heritage

Commercial viability

'The property market's major concerns are with market values and market viability' (Aye *et al.*, 2000). Any additional costs required to develop a sustainable building must be reflected in the benefits available by way of rentals or prices (or reduced operating expenses). So there needs to be a trade-off between capital and revenue costs and benefits in terms of the returns required for viability and the returns likely to be obtained from the market.

Location

Inner-city sites are usually constrained by land area, shape and surrounding buildings that will affect thermal performance and access to natural light. In order to make the best use of solar energy and natural light, the ideal site in the southern hemisphere would have a northern frontage (vice versa in the northern hemisphere) and, with the exception of the western elevation, avoidance of overshadowing by adjoining buildings. Sites in outer suburbs are not so constrained, as cheaper and larger parcels of land are available.

Ownership/tenure

The main types of tenure for commercial office space are either leasehold or freehold. The ownership of business premises provides security of tenure and an associated investment opportunity. However, it ties up capital that might otherwise be applied to the business concern and it requires the owner to face the risks of obsolescence. The leasehold provides all of the rights of ownership but for a limited term. However, the ability of a tenant to obtain high environmental and ecological standards in leased premises is entirely dependent upon a sympathetic landlord. On the other hand, successful sustainable operation of a building is dependent upon the behaviour of the occupants. This could be effected by a lease document that states the operational procedures to be followed by building occupants.

Planning issues

Successful environmental management must also deal with the significant planning issues in which public participation has become so important. The first of these is the concept of appropriate development in which a property development proposal is required by local property owners to be sympathetic to its surroundings. This is reflected in particular in the streetscape and the avoidance of the problems of overlooking and overshadowing of adjacent properties as well as the creation of an adverse microclimate resulting in local wind downdraughts. Issues of transport and traffic and the consequent congestion, noise and pollution problems are of importance to existing property owners and the environment. In addition, the protection of heritage buildings, precincts and streetscapes is a major issue.

Another major environmental concern is the dichotomy between urban consolidation and urban sprawl. One argument is that it is preferable to develop redundant and unused land in urban built-up areas, such as brownfield sites, than to add development to the urban fringe on greenfield sites. However, the capacity of existing infrastructure limits an increase in development density whilst on the other hand new infrastructure would be required for urban expansion. This will affect on an urban scale water supply systems, food production systems, waste management systems, noise pollution and air pollution, energy consumption and the transport system both public and private (Troy, 1996).

A car-parking policy should be formulated, having regard to the dichotomy between market acceptance of the office accommodation and environmental best practice in which all (or at least many of) the building occupants and visitors would access the workplace by public transport.

The foregoing discussion of the environmental issues provides a checklist of features that require decisions during the inception and delivery phase of a project. These features are now discussed in relation to a case study.

9.7   Case study

Several international case studies are now available (National Audubon Society, 1994; Treloar, 1996; Yeang, 1999). This case study comprises a property and construction project, which initially formed the subject of a collaborative research grant, sponsored by the Australian Research Council and industry partners (Bamford, *et al.*, 1997). One of the industry partners had decided to obtain new accommodation for its activities. This entity is at the forefront of environmental matters and wished to occupy premises with a minimum impact on the environment. In the event, an ethical investment company acquired an underdeveloped property and proceeded to commission a building of high environmental design. The decisions required to achieve an environmentally sustainable development are outlined below.

Several key property and construction issues are identified in this study. Many of these issues relate to the acceptance of the project by the property market.

Land and planning issues

Location

In respect of the location of the proposed accommodation, it was decided that the building should ideally be centrally located on a site with excellent public transport access. Accordingly, the search for a suitable location is concentrated in localities within walking distance (a few hundred metres) of the Melbourne underground rail loop which serves the central business district, i.e. in a city fringe locality.

Transport and energy

Key issues here were to optimise the use of public transport and minimise energy consumption and greenhouse gas emissions. Car-parking was not to be provided on the site in an effort to encourage the building occupants to travel by public transport.

Tenure

A preference for freehold ownership was articulated so that the ethical investor has full control of the building design, configuration and operation, and the opportunity to achieve the benefits of sustainable property development and investment.

Embodied energy

Due to the centrality of the chosen location, the ability to find a suitable vacant site was limited. Thus the decision was made to redevelop a site containing an existing building. This decision makes a significant statement about embodied energy and natural resources in that the recycling of (part of) an existing building would create a substantial saving in embodied energy and resource consumption. In turn, this reduced the amount of waste that would need to be deposited as landfill elsewhere.

Heritage

The design includes the preservation of a heritage building and street façade. It is to be incorporated into an office building of modern design. The building has had many alterations and additions during its lifetime and thus has low heritage significance described as remnant architectural merit. Of greater significance is the historical value of the site on which many different activities have taken place. These many past uses justify the changes currently proposed.

Design issues

Passive design

The existing building has three levels and a fourth level is to be added in the form of a lightweight but fully insulated structure. The subject property is constrained by boundary walls on the north and south, a frontage facing west and a rear alignment to a laneway to the east. Therefore there is very little opportunity for natural light and to use the façade as a source of energy. The features used to provide additional light include a central atrium with solar chimneys to effect ventilation and a series of light and ventilation wells on the northern and southern boundaries to provide light to the lower floors of the building and to assist ventilation throughout. Openings equivalent to 2% of the floor area of the building in the external walls and in the roofs of the solar chimneys are required for effective operation of the passive ventilation system. These openings are to be operated in a night purge system to remove heat that has been built up during the day.

Other aspects of passive design include a reinforced concrete structure to provide thermal mass, insulation of all walls and the roof space, low emissivity double glazing, passive solar shading and high-efficiency artificial lighting. Each of these measures taken alone provides a marginal improvement in comfort and energy consumption. All together, these measures should provide commercial quality comfort with a substantial reduction in energy use.

Active design

Given the setting of the property in the commercial office market, comfort conditions associated with conventional office buildings have set the design parameters. The proposed plant is an air-cooled heat pump split system. The system is to be operated on demand when the ambient temperatures exceed the design range (19 to 26°C). The control system will enable the switching over from the passive to the active system and it will be done on a multi-zone basis. In other words, zones requiring specific services are to be operated individually. The greater the number of zones, the greater the potential of energy savings consequent with comfort. The control system will also provide for pre-heating and pre-cooling as necessary.

Transportation services

Although an elevator is provided to enable universal access, building occupants are to be encouraged to use open stairways located in the central atrium.

Electrical services

Connection to the national power supply grid is required. Renewable energy is to be supplied by the retailer (at an additional 33% in price). Limited on-site photovoltaic generation is to be included. PVA sheathed cabling is to be avoided due to its poor fire safety performance. The power density allowance is reduced to 25% of the normal allowance for a building of this type.

Hydraulic services

It is proposed to reuse stormwater on the site but with an allowance for excess stormwater to be reticulated into the infrastructure. Grey water and black water are to be treated on the site and effluent used on landscaping in the atrium and on the roof garden.

Fire services

Environmentally sustainable features are provided in the fire services system. Photo-optical smoke detectors are to be installed instead of the usual ionisation-type detectors to avoid radioactive materials on the site. The backup power supply for the emergency lighting is to be fed by solar collectors. The water used in the testing of the sprinkler system is to be discharged into the grey water system.

Acoustics

It is conceded that the nature of the atrium and the light courts, together with the existing timber floors in parts of the existing building, mitigate against optimum acoustic conditions. Measures are to be put in place to minimise the acoustic impact between the occupants of the building.

Materials

Ecologically sustainable materials are specified including high-mass masonry for the main perimeter walls and lightweight construction in the walls of the atrium and light-wells. The former are to be reinforced concrete or brickwork, some of which is existing, and the latter are to be a mixture of glazing and polycarbonate with perforated pressed metal panels and compressed sheet linings. Floors are to be a mixture of polished concrete, natural timber, carpet and vinyl. Internal partitions are to be glazed or finished in plasterboard. Suspended ceilings are minimised to gain access to the thermal mass.

Energy and greenhouse gas emissions

The building configuration and the land and planning issues and design issues establish the life-cycle energy use and greenhouse gas emissions. The design details have been adopted in an attempt to optimise the sustainability issues together with the economic and financial issues.

Building delivery

Pre-qualification

The consultants and the building contractor have been selected on the basis of a two-stage process. The first stage consists of a public advertisement requesting expressions of interest from suitably qualified persons or firms to present credentials for consideration by the developer. The factor considered important for the first stage is previous experience, particularly in environmentally sustainable development. The second stage commences with the preparation of a short-list of potential contributors and an invitation to submit a proposal including a scope of services to be provided together with the proposed fee in the case of the consultants or the tender price in the case of the contractors. This stage includes an interview of the firms that submitted a proposal in order to meet the individuals who would be involved in the project and to assess their commitment to the nature of the project. Fee and price are not necessarily the major drivers of the selection of the successful consultants and contractors.

Construction operations

The consultants and the contractor are expected to apply sustainable practices to their professional and commercial activities. For example, the consultants are to design and document the project using electronic storage, retrieval and transfer of information. The contractor is expected to provide an environmental management process to enable sustainable site practices. An environmental auditor is to be appointed to advise and be the final arbiter on the project during the delivery phase and the superintendent is to consult the auditor and accept the auditor's advice. The contract is based on an industry standard (Standards Association of Australia, 1992) with the addition of special conditions to reflect the environmental ethic.

Building management

Lease terms and building rules

The building occupants are to be signatories to a lease document that contains requirements, including an environmental management plan and building rules, for the promotion of environmentally sustainable practices. Tenants are expected to operate the building systems for best environmental outcomes. This includes a requirement to commute to the premises by public transport where possible, thus keeping vehicular use to a minimum. As has been discussed above, car-parking spaces on the property are not provided.

Performance monitoring

The building owners intend to monitor the building to establish that the best possible environmental practice is maintained. This will entail the measurement and recording of basic climatic data including temperature, humidity, wind speed and direction, solar radiation and rainfall. A self-contained weather station is to be installed. The building monitoring will also include the measurement and recording of internal data such as temperatures, humidity, electricity consumption, water consumption, lighting levels and so on. A building automation system is to be installed together with logging devices to monitor the operation of the building, in particular the zoned air-conditioning system that is to be controlled by the building occupants and the ventilation controls.

9.8 Conclusion

The conclusion to be drawn from a discussion of the environmental management issues is that property development and building construction can play a significant part in the quest for environmentally/ecologically sustainable development. Many measures are available for reducing the impact on resources, in particular natural resources such as energy, water, air and natural building materials. Contamination and pollution both internally and externally can be optimised. However, continuing legislation and discussion of policy issues is required for improvement in environmental outcomes. The adoption by many countries of international standards on environmental management should lead to an improved performance by the property and construction sector.

The rapid growth of information technology provides opportunities for the management of building performance in the environmental context as well as the financial and economic context.

Greater public participation in these issues will see the stakeholders take greater ownership of their activities. The stakeholders include property owners and occupants, consultants, contractors, the community, neighbours, legislators and future generations.

The new ethic is permeating across the built environment so that decisions are beginning to be considered which are environmentally effective first and cost-effective second. However, sight must not be lost of the commercial aspect of property management. All stakeholders must play their part in the improvement of the environment. The community cannot expect property owners to bear all of the responsibility. This may mean that building occupation becomes a higher-cost activity as the costs of dealing with at least some of the environmental consequences are absorbed.

9.9 Acknowledgements

The author acknowledges the contribution of the Australian Research Council, the industry partners, colleagues at the University of Melbourne and consultants advising on the project outlined in the case study.

References

Australian Greenhouse Office (1991) *Scoping Study for Minimum Energy Performance Requirements for Incorporation into the Building Code of Australia.* AGO.

Aye, L., Bamford, N., Charters, W. & Robinson, J. (1999) Optimising embodied energy in commercial office development. Paper given at *COBRA 1–2 September 1999 – The Challenge of Change: Construction and Building for the Millennium,* University of Salford and Royal Institution of Chartered Surveyors, pp. 217–23.

Aye, L., Bamford, N., Charters, W. & Robinson, J. (2000) Environmentally sustainable development: a life cycle costing approach for a commercial office building in Melbourne Australia. *Construction Management and Economics,* **18**, pp. 927–34.

Bamford, N., Charters, W., Lacey, R. *et al.* (1997) Environment and sustainability in commercial office buildings. Paper given at the *14th Annual Conference, Association of Researchers in Construction Management,* Reading, 1998, 644–51.

Brown, S.K. (1997) Indoor air pollutants: sources, health effects and measurement. Paper given at *Managing Indoor Air Quality,* Australian Insitute of Refrigeration, Air Conditioning and Heating, Melbourne.

Building Owners and Managers Association (now the Property Council of Australia) (1994) *Energy Guidelines.* BOMA.

Building Owners and Managers Association (now the Property Council of Australia) (1996) *Survey of Office Building Energy Consumption.* BOMA.

Calvert, R.E., Bailey, G. & Coles, D. (1995) *Introduction to Building Management*, 6th edn. Newnes.

Cooper, I. (1996) Emerging issues in environmental management. In: *Facilities Management: Theory and Practice* (ed. K. Alexander). E & FN Spon.

Evans, D. (1992) *Transport and Greenhouse: what won't work and what might work*. Papers of the Australasian Transport Research Forum.

Fay, M.R. (1999) *Comparative life cycle energy studies of typical Australian suburban dwellings*. Unpublished PhD thesis.

Flanagan, R. & Norman, G. (1984) *Life Cycle Costing for Construction*. The Royal Institution of Chartered Surveyors.

Fryer, B. (1997) The Practice of Construction Management, 3rd edn. Blackwell Science.

Griffith, A. (1994) Environmental Management in Construction. Macmillan.

Jacques, R. (1996) Energy Efficiency Building Standards Project – review of embodied energy. Paper given at *Embodied Energy: The Current State of Play Seminar* (eds G. Treloar, R. Fay & S. Tucker), Geelong, Deakin University.

Kirschbaum, M.U.F. (1996) The carbon sequestration potential of tree plantation in Australia. In: *Environmental Management: The Role of Euclypts and other Fast Growing Species* (eds K.G. Eldrige, M.P. Crowe & K.M. Old). CSIRO Forestry and Forest Products.

National Audubon Society (1994) *Audubon House: Building the Environmentally Responsible, Energy-efficient Office*. John Wiley.

Prior, J.J. (1993) *BREEAM, Building Research Establishment Environmental Assessment Method: an environmental assessment for new office designs*. Building Research Establishment.

Robinson, J. (1996a) *Life Cycle Costing in Buildings: a practical approach*. Australian Institute of Building Papers. vol. 1, 13–28.

Robinson, J. (1996b) Plant and equipment acquisition: a life cycle costing case study. *Facilities*, **14**(5/6), 21–25.

Scott, S. (2000) A Shrinking Threat. *Australian Financial Review*, 14 Sept. 2000, p. 34.

Standards Association of Australia (1992) *General Conditions of Contract*. Standards Australia, AS-2124.

Standards Australia & Standards New Zealand (1996) *Environmental Management Systems – Specification with Guidance for Use*. Standards Australia.

Standards Australia & Standards New Zealand (1998) *Environmental Management – Life Cycle Assessment – Principles and Framework*. Standards Australia, AS ISO 14040:2.

Standards Australia & Standards New Zealand (1999) *Environmental Management – Vocabulary*. Standards Australia, AS ISO 14050:3.

Stone, P.A. (1980) *Building Design Evaluation: Costs-in-use*, 3rd edn. E & FN Spon.

Treloar, G. (1996) *The Environmental Impact of Construction: a Case Study*. Australia and New Zealand Architectural Science Association.

Troy, P. (1996) *The Perils of Urban Consolidation*. The Federation Press.

Williams, P. (2000) Indoor air quality management: an Australian perspective. Paper given at *IAQ2000 Conference*, Singapore, 11 October.

Yeang, K. (1999) *The Green Skyscraper*. Prestel.

10 Procurement Strategies

Roy Morledge

10.1 Introduction

Chapter 2 refers to the fact that the need for a construction project is rooted in the business case for the project which will be developed by the client. Usually the client will see the new project as a solution to the challenge set by the business case and will need to decide quickly whether the project is viable in terms of its long-term 'worth' and achievable in terms of its initial cost and planned duration.

This concept of long-term 'worth' is wider than capital value and is unique to each client whether individual or corporate. It adopts the concepts of functionality, real estate value and return on investment; often all three. For example, if a project is needed to improve the efficiency of a client's production processes then its worth may be initially rooted in its function. Its real estate value will also add to the company's fixed assets on its balance sheet. If the project has investment potential then the client may also be able to realise investment growth at some point in the future.

Consequently, a successful construction project outcome can provide increased efficiency, greater borrowing capability and payback – long-term worth. This is not, however, always the purpose of the project. Sometimes it is there to serve a short-term need or to be sold on as a completed development. In these cases short-term value rather than long-term worth may be the primary criterion identified in the business case.

This is the key to the development of an appropriate procurement strategy. The business case should always drive the selection of strategy.

10.2 Procurement strategy

The client will probably make the choice of strategy after taking advice and considering all aspects of alternative proposals and potential solutions. After this initial choice the strategic brief will need to be produced defining the project and the parameters within which it must be realised if it is to meet the factors critical to its success as described in Chapter 3.

The decisions taken at this stage will drive all future decisions and strategy. It is a key but complex stage carried out at a time when both client and consultants wish to progress quickly but when the pace of coordination and synergy of the issues and ideas is dictated by the ability to predict, foresee and design. Often the client will be best advised to appoint a member of his own organisation as 'sponsor' to coordinate the client function from this stage, or to appoint a project manager to overview the whole process where the project is particularly complex.

There is a danger that when the process of defining the project is carried out, the primary needs of the business case can be given less prominence as short-term objectives relating to cost and time are seen as increasingly important to the project team. The definition of the project should clearly set down the client's overall business objectives, and indicate the relative priority of each objective. Many construction projects suffer from poor definition due to inadequate time and thought being given at an early stage. This is often because, as the Construction Industry Board (1997) has suggested, 'there is a sense of urgency fuelled by the desire for an immediate solution'.

Prioritisation of objectives will probably be made by considering the impact upon the business case of failing to achieve that particular objective, such as failing to achieve functionality in the new project, or failing to deliver at the planned time.

Project objectives can usually be reduced to the relative importance of the primary criteria of performance (including function and quality), time (including speed and time certainty) and cost (including price level and cost certainty). Prioritisation of these at an early stage will establish a clear policy upon which strategy can be built, providing their prioritisation is linked to the business case.

By considering three examples of types of client and making general assumptions about their likely objectives, the issue of prioritisation can be illustrated (see Table 10.1). By adopting a technique of distributing priorities within parameters (in this case percentage points) the natural tendency to want performance, time and cost to all be high priority is countered. These criteria are interdependent and none of the currently adopted procurement strategies will facilitate the delivery of all of them as high priority. Consequently, prioritisation is necessary. Whilst fast-track solutions deliver speed, other criteria such as cost certainty and functionality

Table 10.1   Examples of prioritised criteria by client type.

	Owner-occupier (%)	Developer (%)	Investor (%)
Performance (functionality/quality)	45	20	50
Time (certainty or speed)	25	50	30
Cost (certainty or price)	30	30	20
Total	100	100	100

may be less achievable, and similarly where cost certainty or performance are considered to be of highest priority the other criteria are affected. Consequently, the distribution of prioritisation within parameters is a useful tool.

It is a reasonable precept that the greater concentration on defining the project and the linking of the prioritisation of objectives to the business case, the greater is the likelihood of the client's objectives being achieved.

Designing and constructing a new building is rarely straightforward. It is subject to a series of risks and uncertainties and involves a number of organisations especially assembled for the project. The way in which the client and the various designers, contractors and suppliers work together as a team is determined by the procurement strategy and the forms of contract entered into between the project participants and the client.

Procurement strategy is the outcome of the series of decisions which are made during the early stages of the project. It is one of the most important decisions facing the client. The chosen strategy influences the allocation of risk, the design strategy and the manner of employment of consultants and contractors. Procurement strategy also has a major impact on the time-scale and ultimate cost of the project.

Generally, clients can choose from several different strategies. Strategies requiring the prioritisation of performance will focus more heavily upon design and will usually require the early appointment of consultant designers. Strategies emphasising speed and/or capital cost may or may not require the involvement of directly appointed design consultants.

Perhaps the key factor in establishing the final strategy is a balance between prioritised objectives and attitude to risk. Figure 10.1 illustrates, in flowchart format, the development of procurement strategy.

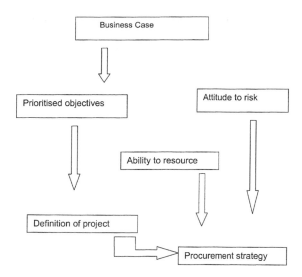

Fig. 10.1   Development of procurement strategy.

10.3   Client risk in construction projects

There is a finite amount of risk and responsibility associated with any project. Risks are usually considered as uncertain future events, which may have significant effects, e.g. extra cost, delay or damage to the performance of the finished project. This section briefly addresses risk and the client in the context of procurement; for a full description of risk management see Chapter 6.

Having set the priorities for the project's objectives, the effect of those objectives not being met and the resulting risks to which the client could be exposed should be considered. Although a project is subject to a wide variety of risks, it is important to note that relatively few have a major effect upon the client. This is a powerful argument for concentrating attention during cost estimating and management decision-making on the few largest sources of uncertainty and risk and for developing strategies for managing out risk and for setting up contracts in such a way that the allocation of the major risks is clear.

Client risks, which are considered to have potentially the greatest impact on construction projects, include:

- A project which will not function in accordance with the client's needs
- A project which is of inadequate quality
- A project which is completed later than required deadlines

- A project which costs more than the client's budget or ability to pay

In each case, the strategy can be to transfer the whole risk to another through the medium of contract. This is possible but will attract high price premiums or will expose the other party to risks which they may not be able to 'own' or insure and therefore the party transferring the risk will remain exposed. Risk may alternatively be retained in part by the client, or reduced by adequate pre-design or pre-price investigation.

In practice, risk allocation is determined by the chosen strategy and allocated by means of contracts between the client and those responsible for managing, designing or constructing the project. Ideally, risk and responsibility should go together, so that the party responsible for performing a task is accountable. Each risk should be allocated to the party with the greatest ability to own the risk and to manage its effects. Responsibility for risk and the ability to control a project interact. The more the client chooses to allocate risk to other parties, the less control the client has over the way in which the project is executed. For example, in a traditional lump sum project where the contractor has to meet the agreed specification within the price and time agreed, the client has little influence over the way in which these objectives are met. If this is further extended to give the contractor liability also for design then the potential for the loss of more control is inherent unless careful steps are taken.

An adequate brief will ensure functionality and quality standards; equally, adequacy of programme will reduce the risk of overrun, and adequacy of cost estimates should ensure a resultant cost, which is within budget. Both construction time and construction cost estimates depend upon sufficient design development, which itself will depend upon an adequate brief and parallel investigation of ground conditions and the particular requirements of statutory controls. In ensuring that the brief is adequate, and that design is appropriately developed, the client can successfully reduce some risk in a way that will not result in high price premiums.

It is inevitable that risks are inherent in the data used by the client in the preparation of the brief; they are inherent in the characteristics of the project; and they are inherent in the procurement strategy that is selected. The identification of primary risks and an analysis of the client's ability to be a risk-taker, or need to be risk-averse, should affect procurement strategy. With this information in mind, steps should be taken to:

- Inform the client of the risks involved
- Prepare a strategy for managing risk

Since the latter will have a cost, the client should also be advised of the likely expense of managing risk. The characteristics of each procurement strategy establish the levels of risk associated with time, cost and performance in each case on the assumption that the procurement strategy is properly utilised and in no way abused. (Thus, for example, in relation to the traditional system it has been assumed that design completion is achieved before measurement and documentation is carried out.) The Construction Clients' Forum (1998) emphasised that for clients 'the avoidance of unpleasant surprises is everything' and that consequently they should be made aware of the nature of the risks to which they are exposed and the likelihood of the occurrence of them.

10.4   Selecting a procurement strategy

Clients who can be described as experienced will be able to select a procurement approach which has worked for them before or which they know to be suitable taking into account their prioritised objectives and their attitude to risk. Inexperienced clients (the majority of clients in the UK fall into this category) will need to seek advice from experienced professionals to help them through the process. It is most important that the strategy is reviewed at key times in the progress of the project such as when planning approval is given, before the contract strategy has been finally decided and before the construction contract(s) has (have) been let. Reasons for this include: (1) the development of design not maintaining the pace anticipated; or (2) the programme is otherwise affected by unexpected occurrences.

The selection of an appropriate procurement strategy has two components:

- Analysis – assessing and setting the priorities of the project objectives and client attitude to risk.
- Choice – considering possible options, evaluating them and selecting the most appropriate.

It may be necessary to seek specialist advice from other consultants, for example in relation to expected costs for the project. Specialist advice should always be sought when developing the strategy for novel or especially difficult projects.

The factors listed below should be considered in analysing project objectives, requirements and their relative priorities (each is then considered in detail):

- Factors outside the control of the project team
- Client resources
- Project characteristics
- Ability to make changes
- Risk management
- Cost issues
- Timing
- Quality and performance

Inevitably, some of these requirements will be in conflict and priorities need to be decided. The choice of strategy should ensure that control is maintained over those factors that are of most importance to the client.

Factors outside the control of the project team

Consideration must be given to economic, commercial, technological, social, political and legal factors that influence the client and the project team or are likely to do so during the lifetime of the project. These may include forecast and actual:

- Interest rates
- Inflation
- Changes in the demand for construction affecting tender price levels
- Legislation

Client resources

The client's knowledge, the experience of the client company's organisation and the environment in which it operates are vital in assessing the appropriate procurement strategy. It is the nature and culture of the client organisation, external influences and the expectations of individuals affected by the project, which influence project objectives. The extent to which the client is prepared to take a full and active role is a major consideration.

Project characteristics

The size, complexity and location of the project should be carefully considered and particular attention given to projects with novel elements. For example, if the building is especially large or complex there may be a bigger risk of cost or time overrun. Novel projects present special risks. The novelty potential factor means that estimates of time, cost and performance are all subject to the possibility of greater error with an increased risk of one or more of the project's objectives failing.

Ability to make changes

It is preferable to identify the total needs of the project during the early stages but this is not always possible. Rapidly changing technology often means late changes. Changes in the scope of the project very often result in increased costs, especially if they arise during construction. Changes introduced after the design is well-advanced or construction has commenced often have a disproportionate effect on the project, in terms of cost, delay and disruption, compared with the change itself. The design process goes through a progressive series of 'freezes' as it develops but the client or project team should set a final design freeze date after which no significant changes to requirements or design are allowed. Some procurement strategies are better than others at handling the introduction of changes later in the project and reducing the possibility of having to pay some form of specific premium.

Cost issues

The cost issues are:

- Need for price certainty. This influences the project timing and the procurement strategy to be used. Generally, design should be complete if price certainty is required before construction commences, unless design-and-build strategies are adopted.
- Impact of inflation where increased price clauses exist.
- Cost of changes. If cost certainty is to be maintained during the course of construction, changes should be avoided. Changes often have cost and time implications on the project well in excess of the change itself. It is therefore important for the client to fix a date after which no significant changes should be introduced.

The total project time frame

Most projects are needed within a time frame or by a specific date. Timing will influence whether subsequent activity can occur as planned and in many cases may severely affect those factors identified as critical or high priority in the business case.

Self-evidently setting unachievable programmes will result in overruns and the UK construction industry has a reputation for delivering projects 'late'. Overoptimism or lack of reliable data on which to rely can be the cause. Realism may frustrate an impatient client but unexpected lateness may have more severe consequences.

The programme of the project is influenced by many of the factors identified in the above section and a particularly large and/or complex project is likely to require more time for design, specification and construction than would a simple small building.

It is of vital importance to allow for adequate design time in terms of the total project. This is particularly the case if design is required to be complete before construction commences (where perhaps cost certainty is required). In the process of the appointment of the design team, assurances should be obtained about resource levels and the ability to meet key dates or programmes. It is not usual to impose contractual dates upon designers, although their progress is probably the key to the overall completion date. Decisions to progress with a project may be influenced by the gaining of planning approval, by the successful operation of a compulsory purchase order, by land purchase or by some other non-specific but critical factor (such as obtaining funding approval). Depending on whether these factors occur earlier or later, they may be an influence upon the planned or desired time available for design.

Some procurement strategies enable an overlap between the design and construction stages, so construction can start earlier than sequential strategies and offer the potential for earlier completion. It may be necessary to review planned procurement strategy in the light of design progress at the point where restraints to constructions are removed, bearing in mind the stage of design and the consequence in terms of risk.

Time has both a cost and a value. If the 'worth' of a project is identifiable then the cost of relative late completion and the value of relatively early completion can be assessed and may form an important factor in the decision-making process. This is often referred to as time/cost trade-off and for relatively early completion may encompass early income flow from a commercial market and will enable reduced interest and insurance charges to be realised. Relatively late completion will attract greater interest and insurance charges (amongst others) and potential loss of opportunity. Developing calculations which identify these sums can be a useful management tool.

Construction time

Total construction time is a consequence of design. More complex structures will almost certainly take longer given the same cost or size, and may require more resources. Although it is possible to work on site for extensive hours or to increase resources, it is not always possible to achieve directly resulting productivity. The law of diminishing returns will have an influence because of the limited space and the nature of traditional construction methods (such as concreting and bricklaying).

Performance

The required performance of the project measured both in terms of its response to the needs of the client as expressed in the business case and the quality of individual elements should be clearly identified. If performance is over-specified, a premium will be paid for exceeding actual requirements, thereby affecting the cost objective. Over-specification will also lead to time overruns. Conversely, failure to recognise the true performance objective leads to an unsatisfactory product. If quality and performance are particularly important the client will probably want to keep direct control over the development of the design. This can be achieved by employing the design team directly.

10.5   Procurement strategy selection checklists

These checklists have been designed to establish the range of information about client needs and about the particular project being considered, as discussed above, and to develop this information in parallel with the characteristics of procurement strategies and associated risk. They are intended to inform judgement, not to replace judgement. The relative importance of time, cost and performance (design) form key criteria in the selection mechanism, as does inherent risk and its apportionment.

Method

Checklists 1, 2 and 3 should be carefully completed in consultation with the client. The resultant information should then be transferred to checklist 4, which will enable the information to be analysed and evaluated. The analysis can then be compared with the characteristics of each procurement method, and the associated risk so that an informed decision can finally be made.

Checklist 1: Time

The following should be considered:

1.1 Is completion needed by a specific date?
1.2 Is completion needed in the shortest possible time?
1.3 Is the client prepared to pay more for earlier completion?

2. Does the answer to question 1.1 suggest a faster than 'normal' total project time in the judgement of the client's advisers?

3. How long is it in months from the date of completion of this protocol until the desired 'move in' date?

4. Define the reason for the identified completion or 'move in' date:
   4.1 End of lease
   4.2 Sale of premises
   4.3 New business opportunities
   4.4 Unsuitability of present premises
   4.5 Company restructuring
   4.6 Other

5.1 Is the need for completion by a specific date or within a specific time more important than certainty of construction cost before work starts?

5.2 Is the need for completion by a specific date or within a specific time more important than spending an extended time on design?

6. What is the approximate monetary value to the client of the building or facility in terms of contribution, rental or cost savings per month?

7. If the building is completed later than the specified or desired time, will the client:
   7.1 Stay in his existing premises?
   7.2 Find temporary accommodation?
   7.3 Close down?
   7.4 Adopt some other course?

Upon completion of this part, the following information should have been established and should be transferred to checklist 4:

1a Specified completion time
1b Reason for completion time
1c Whether required completion time is relatively fast
1d Whether time is seen as the predominant client need
1e The potential financial implication of earlier or later completion
1f What action the client may take if the dates are not achieved

Checklist 2: Design/performance

1. Has the client clear ideas about building functionality and its desired design?

2. Does the site (if selected) pose any particular problems for the designer in respect of:
   2.1 Shape or topography?
   2.2 Access?
   2.3 Storage space?
   2.4 Contamination?

3. Does the building type suggest relative design complexity?

4. Does the building type suggest emphasis upon functionality?

5. Does the building type suggest highly complex mechanical, electrical or engineering installations?

6. Is it anticipated that extensive changes to design may be required during the construction phase?

7. Does the client wish to emphasise low running costs?

8. Does the client wish to emphasise low maintenance costs?

9. Does the client wish to emphasise product quality at a higher potential cost?

Upon completion of this part of the process, the following information should have been established and should be transferred to checklist 4:

2a Whether the client has clear ideas about his needs
2b Whether the site poses complex design problems
2c Whether the building design is complex
2d Whether functionality is particularly important
2e Whether the client has a long-term view about the cost of the building

Checklist 3: Cost

1. What is the client's maximum budget?

2. Can the budget be allocated as below?
   2.1 Land purchase and fees
   2.2 Construction, including fees
   2.3 Fittings and plant
   2.4 Contingencies
   2.5 Other (define)

3. Will the client need to have a fixed contract price for the construction element of his budget or will a reasonably accurate budget be adequate?

Upon completion of this part of the process, the following information should have been established and should be transferred to checklist 4:

3a   Total maximum spending capacity
3b   Total construction spending capacity
3c   Need for pre-construction cost certainty

Checklist 4: Analysis

As an overview, is the project feasible in terms of time and viable in terms of cost on the basis of the information in 1a, 2a, 2b, 2c, 3a, 3b, 3c?
If yes, proceed.
If no, advise client and seek other solutions.

Is the reason given as to why it should be completed by the specified date vital to the project's success in terms of client needs, or can slippage be coped with if cost is considered?
See 1b, 1d, 2a, 2b, 2c, 3.
If vital, the project has to be carefully planned from a reasonably advanced design. Design and construction must be planned. Where design is completed, the completion date can be contractually fixed. Where design cannot be completed, fast-track systems can achieve relative speed by overlapping design and construction.

If relative speed is required, can the client accept less cost certainty?
See 1c, 3c.
If yes, this may mean that fast-track approaches may be suitable.
If no, a method of achieving cost certainty relatively quickly may be through negotiation.

Does the information provided indicate that the project is complex in terms of design or in terms of site-related problems?
See 2b, 2c, 2d.
If so, adequate time must be allowed for a design process to occur which will provide the client with an acceptable design solution.
Some fast-track solutions will enable the design process to be extended into the construction phase where pre-construction time is not available. This will reduce cost certainty.

Is the need to ensure that the project can be built within a budget a priority?
See 3c.
In this event, design should be complete before construction is

commenced or a sufficiently large contingency should be allowed. The latter will not give cost certainty and may result in a less balanced project in financial terms.

Checklist summary

Having established whether the project is both feasible and viable, the importance of time, cost and design has now been reviewed. There is always an interrelationship between these three primary criteria and procurement systems selected should reflect this. Projects can probably be broadly categorised into those which will be design-led and those which will be production-led. Design-led projects usually reflect the characteristics of procurement systems where the client appoints his own design team, whereas production-led projects enable the constructor to take on some or all of the design function. Design-led projects have the greater capacity for cost and time overrun, whereas the potential capacity for design shortcomings may rest with product-led systems.

System selection should consider whether the project is design- or product-led but should also consider the client's need to manage and/or distribute risk.

10.6   Procurement options

When all the factors influencing the project have been identified and the project requirements analysed, the final strategy for the project must be developed.

It is likely that there will be more than one way to successfully achieve the requirements of the project. It is important to consider carefully each option, as each will address the various influencing factors to a different extent. In developing strategies, a potential danger is that only the most obvious course of action may be considered; this is not necessarily the best in the longer term.

Traditional (design–bid–build)

Under a traditional procurement strategy, design should be completed before tenders are invited and the main construction contract is let. As a result, and assuming no changes are introduced, construction costs can be determined with reasonable certainty before construction starts.

The contractor assumes responsibility and financial risks for the building works whilst the client takes the responsibility and risk for design team performance. Therefore, if the contractor's works are delayed by the failure of the design team to meet their obligations, the contractor may seek recompense from the client for additional costs and/or time to complete the project. In turn, the client could seek to recover these costs from the design team members responsible, if negligence could be proved.

Clients are able to influence the development of the design to meet their requirements because they have direct contractual relationships with the design team. When construction begins, they usually have a single contractual relationship with a main contractor and, therefore, are able to influence (but not control) the construction process through a single point of contact.

The strategy may fail to some degree if any attempt is made to let the work before the design is complete. Such action will probably result in many post-contract changes which could delay the progress of the works and increase the costs. Where a traditional strategy is chosen because of its particular advantages, the temptation to let the work before the design is complete should be strongly resisted.

It is possible to have an accelerated traditional procurement strategy where some design overlaps construction. This can be achieved by letting a separate, advance works contract, for example by allowing ground works (site clearance, piling and foundations) to proceed to construction whilst the design for the rest of the building is completed, and the construction tendered separately. This reduces the total time to complete the project at the risk of losing certainty of cost before construction starts. More importantly, a substantial risk is created in that the contractor who builds the superstructure has no responsibility for the foundation works carried out by another contractor.

Another option is to let the work by a two-stage process or by negotiation, thus reducing the pre-construction time involved. Standard forms of contract are available.

The main advantages of a traditional contract strategy are:

- Competitive fairness
- Design-led, facilitating high level of quality in design
- Reasonable price certainty at contract award based upon market forces
- The strategy is satisfactory in terms of public accountability
- The procedure is well known
- Changes are reasonably easy to arrange and value

The main disadvantages are:

- The strategy is open to abuse, resulting in less certainty
- The overall programme may be longer than for other strategies as there is no parallel working

- No 'buildability' input by contractor
- The strategy can result in adversarial relationships developing.

Measurement contracts

A measurement contract is where the contract sum is only established with certainty on completion of construction, when remeasurement of the quantities of work actually carried out takes place and is then valued on an agreed basis. Measurement contracts are sometimes referred to as remeasurement contracts and are based upon the prices tendered by the contractor. They are used when the work required cannot be accurately measured for the tender bill of quantities. The most effective use of a measurement contract is where the work has been substantially designed but final detail has not been completed. Here, a tender based on drawings and a bill of approximate quantities will be satisfactory. Measurement contracts allow a client to shorten the overall programme for design, tendering and construction but usually with the result of some lack of price certainty at contract stage because the approximate quantities reflect the lack of information on exactly what is to be built at tender stage. The scope of the work, the approximate price and a programme should be clear at contract stage. Measurement contracts provide more risk than lump sum contracts for the client but probably with programme advantages. Standard forms of contract are available for this strategy.

The main advantages of the measurement contract strategy include:

- Pre-construction time-saving potential
- Competitive prices
- Average public accountability
- The procedure is well known
- The strategy enables changes to be made easily
- There is some parallel working possible

The main disadvantages of the strategy are:

- The strategy offers poor certainty of price
- There is no contractor involvement in the planning or design stage
- There is a potential for adversarial relationships to develop

Construction management

Under a construction management strategy, the client does not allocate risk and responsibility to a single main contractor. Instead, the client employs the design team and a construction manager is engaged as a fee-earning professional to programme and coordinate the design and construction activities and to improve the build-ability of the design. Construction work is carried out by trade contractors through direct contracts with the client for distinct trade or work packages. The construction manager supervises the construction process and coordinates this with the design team. The construction manager, who has no contractual links with the design team or the trade contractors, provides professional construction expertise without assuming financial risk, and is liable only for negligence by failing to perform the role with reasonable skill and care, unless some greater liability is incorporated in the contract.

On appointment, the construction manager will take over any preliminary scheduling and costing information already prepared and draw up a detailed programme of pre-construction activities. Key dates are normally inserted at which client decisions will be required. In adopting the construction management system, the client will be closely involved in each stage of design and construction. The client must have administrative or project management staff with the time and ability to assess the recommendations of the construction manager and take the necessary action. The client needs to maintain a strong presence through a project management team that is technically and commercially astute. This strategy is not, therefore, suitable for the inexpert or inexperienced client.

With this contract strategy, design and construction can overlap. As this speeds up the project, construction management is known as a 'fast-track' strategy. Although the time for completion can be reduced, price certainty is not achieved until the design and construction have advanced to the extent that all the construction work (trade) packages have been let. Also, design development of later packages can affect construction work already completed. The construction manager should, therefore, have a good track record in cost forecasting and cost management. A package is made up of work for which one of the trade contractors is responsible, e.g. foundations, concrete, electrical installation or decorating. These packages are tendered individually, for a lump sum price, usually on the basis of drawings and specification.

Construction management has been used predominantly for large and/or complex projects, but there is no intrinsic reason for this. Indeed, it is particularly recommended for projects where there is a high degree of design innovation, where the client wants 'hands on' involvement. There is currently no standard form of construction management contract. Companies who act as construction managers normally use forms they have developed themselves.

The main advantages of a construction management strategy are:

- The strategy offers time-saving potential for overall project time due to the overlapping of procedures.
- Buildability potential is inherent.
- The breakdown of traditional adversarial barriers.
- Parallel working is an inherent feature.
- Clarity of roles, risks and relationships for all participants.
- Changes in design can be accommodated later than some other strategies, without paying a premium, provided the relevant trade packages have not been let and earlier awarded packages are not too adversely affected.
- The client has direct contracts with trade contractors and pays them directly. (There is some evidence that this results in lower prices because of improved cash flow certainty.)

The disadvantages are:

- Price certainty is not achieved until the last trade packages have been let
- An informed, proactive client is required in order to operate such a strategy
- The client must provide a good-quality brief
- The strategy relies upon the client having a good quality team
- Time and information control is required

Management contracting

With this contract strategy, a management contractor is engaged by the client to manage the building process and is paid a fee. The management contractor is responsible for all the construction works and has direct contractual links with all the works contractors. The management contractor, therefore, bears the responsibility for the construction works without actually carrying out any of that work. The management contractor may provide some of the common services on site, such as office accommodation, tower cranes, hoists and security, which are shared by the works contractors. The client employs the design team and, therefore, bears the risk of the design team delaying construction for reasons other than negligence.

Management contracting is a 'fast-track' strategy. All design work will not be complete before the first works contractors start work, although the design necessary for those packages must be complete. As design is completed, subsequent packages of work are tendered and let. Cost certainty is, thus, not achieved until all works contractors have been appointed. A high level of cost management is, therefore, required.

With the agreement of the client, the management contractor

selects works contractors by competitive tender to undertake sections of the construction work. The client reimburses the cost of these work packages to the management contractor who, in turn, pays the works contractors. The management contractor coordinates the release of information from the design team to the works contractors.

Where the management construction team is not of the highest quality, or where this fee is inadequate, the management contractor can be less than proactive and the system can become a reactive 'postbox' approach. It is, therefore, vital to select the management contractor carefully and to ensure that his fee is appropriate, bearing in mind market conditions. Similarly, resistance to works contractors' claims can be affected by the same circumstances. A standard form of contract is available for this strategy.

The main advantages of a management contracting strategy are:

- Time-saving potential for overall project time
- Buildability potential
- Breaks down traditional adversarial barriers
- Parallel working is inherent
- Changes can be accommodated provided packages affected have not been let and there is little or no impact on those already let
- Works packages are let competitively

The disadvantages are:

- The need for a good-quality brief
- Poor certainty of price is offered
- Relies on a good-quality team
- It may become no more than a 'postbox' system in certain circumstances; and reduces resistance to works contractors' claims

Design-and-manage

A design-and-manage strategy is similar to management contracting. Under a design-and-manage contract, the contractor is paid a fee to manage and assume responsibility, not only for the works contractors, but also for the design team.

The main advantages of a design-and-manage strategy are:

- Early completion is possible because of overlapping activities
- The client deals with one firm only
- It can be applied to a complex building
- The contractor assumes the risk and responsibility for the integration of the design with construction

The disadvantages are:

- Price certainty is not achieved until the last work package has been let.
- The client loses direct control over the design quality.
- The client has no direct contractual relationship with the works contractors or the design team and it is, therefore, difficult for the client to recover costs if they fail to meet their obligations.

Design-and-build

Under a design-and-build strategy, a single contractor assumes the risk and responsibility for designing and building the project, in return for a fixed-price lump sum. A variant, known as 'develop and construct', describes the strategy when the client appoints designers to prepare the concept design before the contractor assumes responsibility for completing the detailed design and constructing the works. Design-and-build is a fast-track strategy. Construction can start before all the detailed design is completed, but at the contractor's risk.

By transferring risk to the contractor, the client loses some control over the project. Any client requirement, which is not directly specified in the tender documents, will constitute a change or variation to the contract. Changes are usually more expensive to introduce after the contract has been let, compared with other types of strategy.

It is imperative, therefore, that the brief and performance/quality specifications for important requirements in the project are fully and unambiguously defined before entering into this type of contract. If requirements are not specific, the client should provide contractors with a performance specification at tender stage. The contractor develops the design from the specification, submitting detailed proposals to the client to establish that they are in accordance with the requirements of the specification. Clients are, therefore, in a strong position to ensure that their interpretation of the specification takes preference over the contractor's. Specification is a risky area for inexperienced clients: overspecification can cut out useful specialist experience; underspecification can be exploited.

The client will often employ a design team to carry out some preliminary design and prepare the project brief and other tender documents. Sometimes, the successful contractor will assume responsibility for this design team and use them to produce the detailed design. If a design-and-build strategy is identified as a possibility at an early stage, then the basis of the appointment of the design team should reflect this possibility. If it does not, the client may have to pay a termination fee to the design team. The client may wish to retain the independent services of a cost consultant throughout the contract for early cost advice involvement in the bidding process and cost reporting during construction. Consideration may need to be given to the inclusion of a special clause in design, or design-and-build type, contracts to ensure that the responsibility for design performance is properly shared. Such a clause should identify specific obligations that are absolute, that do not require the designer to be an expert in the client's business and, as a consequence, are reasonably insurable.

Current forms of contract for design-and-build vary their treatment of design liability. Under JCT WCD 98 with Contractor's Design, the contractor undertakes to design to the same standard 'as would an architect if the employer had engaged one direct'. So the client gets the same 'skill and care' liability for design as would normally be available under traditional and management strategies when employing the architect direct.

To be effective, the client's requirements will need to be stated clearly and accurately and delivered on time. The imposition on contractors of fitness-for-purpose in design is a matter of judgement for clients and their professional advisers, even though tenderers in recessionary markets are likely to agree to undertake such risks.

Non-availability of insurance to cover a higher than normal risk should be weighed against the financial ability of contractors to meet design default claims. It will normally be preferable and represent better value for money to impose a lesser, yet insurable, liability, which will be the subject of an insurance payout in the event of a design fault, rather than an insurance fitness-for-purpose requirement on a contractor of limited financial assets. Increasingly, collateral warranties are being used to place additional responsibility on designers or subcontractors. In addition, the client may take out BUILD insurance to cover post-construction liability, but this will be at a cost. This is a matter for specialist advice.

A range of options is available – from a package deal or turnkey where the client has little involvement in the design development or building procurement process (effectively, a complete hands-off approach), to develop-and-construct where the client appoints designers to develop his brief to a level of sophistication which will leave the design-and-build contractor to develop detail or specialist design elements.

Contractors may use their own firms' resources for undertaking the design (in-house design and build), or may subcontract these to one or more professional firms whilst retaining control. A standard form of design-and-build contract is available.

The main advantages of a design-and-build strategy are:

- The client has only to deal with one firm.
- Inherent buildability is achieved.
- Price certainty is obtained before construction starts, provided the client's requirements are adequately specified and changes are not introduced.
- Reduced total project time due to early completion is possible because of overlapping activities.

The disadvantages are:

- Relatively fewer firms offer the design-and-build service so there is less real competition.
- The client is required to commit himself before the detailed designs are completed
- In-house design-and-build firms are an entity, so compensation for weak parts of the firm is not possible.
- There is no design overview unless separate consultants are appointed by the client for this purpose.
- Difficulties can be experienced by clients in preparing an adequate brief.
- Bids are difficult to compare since each design, programme and cost will vary.
- Design liability is limited by the standard contract.
- Client changes to project scope can be expensive.

Procurement Options Summary

Each of these options have their own characteristics in terms of risk, processes and sequence. On some projects it may be necessary to use more than one strategy to meet the project's objectives. For example, a traditional approach may be used for completing the building structurally with main services installed. This is known as a shell-and-core contract. A separate strategy, for example construction management, may be used to fit out the building with ceilings, raised floors, carpets, partitions and electrical fittings. The use of two strategies allows the client more time to finalise the user's detailed requirements, without delaying the start of construction.

When the choice of procurement strategy has been made, the resultant contract strategy and form of contract must be chosen (i.e. the terms and conditions of the contract). To avoid the need for fresh legal drafting each time, various standard forms of contract are available. Most of the standard forms of contract are published by professional bodies in the construction industry and have been developed largely for private sector clients.

Common strategies differ from each other in relation to:

- The financial risk that the client is exposed to.
- The extent and nature of any competition inherent in the adopted strategy.
- The degree of control that the client has over the design and construction processes.
- The information required at the time construction contracts are let.
- The extent of involvement of the contractor in the design stage when the contractor may be able to influence the 'buildability' of the project.
- The organisational arrangements which distribute responsibility and accountability.
- The sequential nature of the process.

10.7   Constructing improvement in procurement strategy

As indicated above, the selection of strategy should be the consequence of the business case and the project definition, taking into account the client's attitude to risk. Traditionally, construction projects have been let on the basis of lowest bid for a defined project. This has been for many years the dominant procurement strategy adopted, and it still is. Surveys of the performance of the industry, the latest and perhaps most important being the CCF Clients' Satisfaction Survey (Construction Clients' Forum, 2000), indicate that there is scope for improvement in the processes adopted in the procurement process and particularly that continuous improvement is most desirable. The CCF survey showed that clients' experience of the construction process had not significantly improved since similar surveys some four to five years ago and that current practices seem to deliver late projects, many over budget with inherent defects and disputes. Clients have stated that they do not like unpleasant surprises (Construction Clients' Forum, 1998) but it seems they are usually party to contracts where such surprises are the norm rather than the exception.

Clients have also asked for continuous measurable improvement (Construction Clients' Forum, 1998), and perhaps collaborative strategies such as partnering can improve each of these areas of poor performance, and hopefully in due time will do so. Partnering has been, and is, held out as one way of achieving process improvement in the UK Construction Industry. Reports including Constructing the Team (commonly referred to as the Latham Report, 1994) and Rethinking Construction (commonly referred to as the Egan Report, 1998) have strongly supported partnering and are beginning to collate some impressive evidence as to the benefits and gains being achieved. Partnering, however, often requires a cultural change in organisations that have for many years been encouraged to adopt a competitive price-focused approach to buying construction and construction services.

Change of any sort is difficult; cultural change, since it goes to the root of an organisation, is particularly difficult. Those who recommend partnering advise that, to be successful, the partnering culture must be driven from the top by chief executives who are themselves committed and who insist that the philosophy is adopted right through their supply chain. It is based upon the concept of the whole project team working towards the same objectives as the client, being allowed a reasonable profit for their efforts and adopting less adversarial attitudes. The partnering philosophy encompasses the development of mutual objectives, an agreed method of dealing with disagreements and continuous measurable improvement. It is rooted in the development of collaboration and trust between parties who have for many years existed in a climate of competitive commercial contracts and adversarialism. This ethos is consistent with many of the philosophies successfully adopted in manufacturing industries where the processes and products are improved on a continuous basis.

Partnering is a people-based philosophy requiring trust and commitment. Instruction by managers can change processes but people need convincing before the philosophy is adopted. Larger, bureaucratic organisations consist of many layers encompassing often-rigid procedures with which people are comfortable. Changing the procedures and the philosophy may take a long time and will depend upon convincing all players, not just key players, that the end is worth the means.

Perhaps all clients can in time focus on fully collaborative strategies such as partnering where the focus is upon achieving best value for the client rather than driving down capital cost, as the present Government is encouraging its own procurers of construction to do. Consequently, a climate more conducive to a wider acceptance of partnering arrangements can grow. In the interim period, however, it may be possible and more palatable to adopt processes that are more easily accepted in the transition but which can improve measured performance sooner.

Where process performance is (as it usually is) measured against budget or programme there are collaborative (part-partnering) strategies and techniques which can be easily adopted without significant variance to established procurement strategies. These techniques involve collaboration between the client's team and the construction team in improving estimates of cost and time and in enabling a level of constructor involvement in the design process.

To enable the effect of the collaboration to be measured, key performance indicators will need to be adopted. These already exist through the Construction Best Practice Programme and have been adopted by the Construction Industry Board as being appropriate in mapping improvement. They accommodate benchmarking of previous cost or time performance as achieved rather than estimated, and a comparison between estimates and performance.

There are a number of established procurement systems that will enable collaboration between the client, his/her consultants and constructors, the extent of the collaboration being dependent upon the strategy adopted. For example, two-stage tendering based upon partly completed design, enables a largely traditionalist competitive approach to contractor selection but facilitates some constructor involvement once the contractor has been selected on the basis of prices against approximate quantities or similar criteria. Constructors can advise on the likelihood of the project meeting its desirable completion date and will be able to assist the design team in terms of the practicability of the detailing. This level of involvement may enable some improvement in the accuracy of time management and level of buildability but may be too late in the process for much to be gained in terms of value for money through design review. Nonetheless more certainty in terms of construction time may be achieved and some benefits associated with establishing critical design activities may ensue.

On the other hand the selection of procurement strategies such as management contracting and construction management facilitates a high level of constructor involvement, enabling design buildability, value management and concept briefing and the benefit from early constructor involvement in terms of estimates of construction programme and cost. The adoption of these strategies is numerically very low, however, when compared with traditional and design-and-build strategies.

Design-and-build strategies anticipate by their very nature collaboration between constructors and the client team. Perhaps this is at its most effective where a develop-and-construct or novated approach is adopted and there can be initial design input from the client's design team rather than just from the contractor. The contractor can then have a level of input in terms of programming and costing, provided the point of novation is not left too late in the process to gain the maximum benefit.

Since most estimates of time and cost against which current performance is judged preclude the involvement of those most able to estimate both (i.e. the constructors), it is argued that their involvement will improve accuracy. After all, it may simply be that current estimates, in lacking this involvement, were actually never achievable and the industry is criticising itself for not achieving impossibility.

Any shift towards adoption of these strategies facilitating collaboration can, however, be quite a step for some organisations. This is usually because to some extent contractual risk shifts back to the client in each case and most clients will usually want to pass risk to others, predominantly the constructor. However, detailed consideration of risk allocation may reveal that by accepting a little more risk through the adoption of collaborative strategies, predictability may be increased and consequently overall risk reduced.

Constructors understand better than designers do the logical progression of construction and how long construction operations take. They also understand construction costs and are able to apply estimating knowledge and techniques not available to the design team. Their involvement will therefore improve predictions and may improve the performance of the whole team in setting achievable targets, achieving better value for the client and improving feedback loops between constructors and designers.

Why constructors should collaborate needs a little more consideration. Where a strategy of full partnering is considered, the constructors benefit by better profitability and predictability of workload. In the case of collaborative strategies the benefits are less obvious. Contractors are themselves, however, significantly disadvantaged by seemingly feeling forced, for business reasons, to attempt to achieve what they believe to be impossible or unlikely targets. Indeed there is evidence that some major contractors are seriously disaffected by the competitive industry, with Cook (1999) stating that half of the top contractors in the UK a decade ago have pulled out of traditional volume work in their core business.

If contractors are set impossible time targets they will strive to achieve them, but when they begin to fail will seek defences from the financial consequences which follow. These defences are frequently successful with extensions of time being commonplace and clients' attempts to manage time-related risks failing. Similarly, if they are beaten down on price to meet an unachievable budget this will initiate a series of actions along the supply chain seeking low bids. It will not achieve a quality focus and will usually result in an adversarial culture.

It is in a contractor's interest for these reasons, and for his/her reputation (increasingly important), to enable realistic targets to be set and achieved by bringing management and estimating skills to a process which usually bases its estimates on a range of second-best data and unreasonable demands.

By involving constructors in a collaborative way, process improvement may be achieved and clients may receive fewer unpleasant surprises. More projects may come in 'on time' and to budget. There are fewer losers and a win–win culture can grow. More importantly, client confidence in the industry can grow. Business builds on confidence.

Collaborative practice is not full partnering but a simple step towards drawing together those elements of a fragmented industry which have traditionally regarded each other with suspicion and mistrust and providing an opportunity for continuous improvement. It does require clients and those who advise them to take a small but bold step and to see what follows as measured against preset indicators, but at little risk, with the potential for significant gain.

Clients must decide whether they will adopt a traditional competitive-bid approach or collaborative or part-collaborative procurement strategies. This may depend upon the amount of leverage or buying power which they hold. Many one-off or inexperienced clients are likely, for now, to continue to adopt traditional strategies.

10.8   Case studies

Three case studies of hypothetical clients and hypothetical projects are used below to illustrate some of the main points raised in the text. These case studies are oversimplified for this purpose and the issues identified can be much more complex and difficult to address. Nonetheless, the checklist does assist in this difficult process.

Case study 1

Acme Food Sales plc is a manufacturer of 'long-life' ready meals, which has grown steadily over the last ten years, reflecting the demand for its product which is supplied to supermarkets nationally. Competition from other manufacturers has now begun to impact upon sales and the directors are concerned that growth and profitability may not be maintained at projected levels.

After extensive discussion the directors believe that they must improve efficiency through a move from their existing premises to an entirely new purpose-built facility, enabling less double-handling and transportation of raw materials through a move to greater mechanisation. They have already targeted a potential site on a nearby industrial estate and the local authority has indicated its general support for the venture. The company is in a position such that funding for the project can be arranged through the company's bank at bank rate + 2%, subject to the provision of projected cash flows and future profit projections. The directors consider that this is an urgent issue and have expressed a wish to complete the move as soon as possible.

When they prioritise their objectives, the directors of this company will probably set performance with the highest priority with cost and time of similar weighting, even though they have expressed a wish for urgency. This is because failure to achieve a new facility which is functionally more efficient would be their greatest risk and would defeat the purpose of their business plan. If speed were considered to be a relatively high second priority because of particular threats to the business, they could adopt a two-stage strategy to increase the pace of the process.

Analysis using the checklist will point to a design-led strategy because of this. Since they are likely to be inexperienced in the purchase of construction, they should be strongly advised to seek the advice of an experienced professional to overview the project and they should appoint an internal 'sponsor' who can coordinate internal decision-making and communication.

Comment

The most appropriate procurement approach for this project is probably to adopt traditional design–bid–build, selecting the consultants and the contractor on the basis of best value.

Case study 2

Flatside Properties plc is a development company whose primary business is the building of regional distribution centres throughout the country to enable the main store chains to have access to local markets for their goods without having to haul them great distances in small vehicles. They identify the sites, buy the land, obtain planning permission, use a local contractor to build the units and then lease them to the store chains on the basis of an annual rental for a lease period of 15 years.

Flatside's primary objectives relate to speed within a limited cost. They are likely to prioritise speed if the market is buoyant for their properties, with cost scoring second and performance last. Their greatest risk is not having lettable properties available at the right time.

Comment

Analysis using the checklist suggests a constructor-led solution and this sort of project probably best suits a competitive design–build solution.

Case study 3

Safe and Sound plc is an insurance company which invests its money in construction projects in the commercial and industrial sector. The company seeks long physical, functional and economic life for its projects as a hedge against inflation and a good long-term investment. Consequently its greatest risk is associated with performance. It will depend upon economic conditions and the demand for commercial property whether speed is key or not.

Comment

Analysis of the checklist will indicate whether or not this is the case, but in any event a design-led solution is suggested. If speed is an issue then a management-based procurement option, such as management contracting and/or construction management could be adopted, the latter requiring a proactive role from an experienced client.

10.9   Conclusion

This chapter has emphasised the importance of linking the business case and the client's attitude to risk with the procurement strategy developed for the project. It has reflected upon the dysfunctionality which can occur between those involved with the development of the business case and those involved with project delivery. Simple steps to avoid such dysfunctionality have been suggested and moves towards achieving improvement through increased collaboration discussed. The achievement of improvement will, however, always depend upon informed judgement and effective project management.

References

Construction Clients' Forum (1998) *Constructing Improvement*. Construction Clients' Forum.

Construction Clients' Forum (2000) *Survey of Construction Clients' Satisfaction 1999/2000*. Construction Clients' Forum.

Construction Industry Board (1997) *Briefing the Team: a guide to better briefing for clients*. Thomas Telford.

Cook, P. (1999) *Building*, November, The Builder Group.

Egan, Sir John (1998) *Rethinking Construction* (the Egan Report). Department of the Environment, Transport and the Regions. HMSO.

Latham, Sir Michael (1994) *Constructing the Team* (The Latham Report). HMSO.

11 Supply Chain Management

Norman Fisher and Roy Morledge

11.1 An introduction

Tough competition and globalisation are two of the drivers which have caused supply chain management techniques to be adopted by industries such as manufacturing and food retailing. Worldwide mergers and acquisitions, amongst for example the world's large retailers, will continue to ensure that supply chain issues remain on the agenda of company managers in these sectors. It is estimated that the globalisation of the manufacturing sector will continue for some time, with companies becoming more international and thus more powerful. The trend towards sourcing a product from a single regional factory will continue, with the initiative moving to the retailer – seen as the organiser of the business idea (identification, creation, distribution, selling process).

As manufacturers and distributor-retailers increasingly operate on a regional or global scale, the networks created will become yet more complex and require much tighter management. For example, a DIY retail chain importing electrical fittings from South Korea to Europe may find distribution more effective from one mainland European location and not from individual country sites. To do this well requires high-quality, reliable information systems, which can transmit changing market requirements quickly and accurately back down the 'extended supply chain' to ensure that obsolete products are not produced. Major retailers, for example Tesco and Wal-Mart, have gained reputations and competitive advantage from this type of supply chain management expertise. Because of all of this, manufacturers and retailers are looking carefully at supply chain management. They see it as perhaps the most effective way of cutting costs and therefore giving sophisticated customers ever-better value. Manufacturers and retailers increasingly see competitive advantage as not being about technology, but rather about supply chain issues. That is about getting the product to where customers want it, at the right price and at the right value level (Fearne, 1996; Batchelor, 1999).

This chapter is a critical review of the literature on supply chain management. It does not attempt a comprehensive description of supply chain management', but rather puts the key principles, benefits and concerns in the context of a complex and fragmented industry.

11.2 Some general definitions

A wide reading of the literature would suggest that supply chain management is rather an umbrella term and when in general use, often includes related topics such as partnering and lean manufacturing.

A supply chain is:

'... a network of facilities and distribution options that performs the functions of procurement of materials, transformation of these materials into intermediate and finished products, and the distribution of these finished products to customers. Supply chains exist in both service and manufacturing organisations, although the complexity of the chain may vary greatly from industry to industry and firm to firm.'

(Ganeshan and Harrison, 2000)

Normal traditional practice has been for company departments such as purchasing, marketing and manufacturing to operate independently of one another, each with their own differing objectives. For example a marketing department with its concerns with sales maximisation and customer/client care could easily be in conflict with a production department's emphasis on production targets and cost reduction. Consequently, there is no single plan that brings the different functions together in an integrated way, with the emphasis on the customer's requirements and on the value received. Supply chain management is a process that enables the required integration to take place and to be managed on a continuous basis. In supporting this view, Ganeshan and Harrison (2000) suggest that supply chain management is typically viewed as lying on a scale between fully vertically integrated firms where the entire material flow is owned by a single firm and those where each product group operates independently. Thus coordination between the various parties in the chain is key to its effective management. Cooper and Ellram (1993) see similarities between supply chain management and a well-balanced and well-practised athletics track relay team. Such a team is more competitive when each of the players knows how to position himself for and deliver the baton transfer. Although the relationships are key between players who directly pass the baton to each other, the entire team needs to make a carefully coordinated effort if they are to win.

Ganeshan and Harrison (2000) describe two categories of decisions for supply chain management: strategic and operational. The essential difference is the time horizon over which the decisions are made. Strategic decisions, as one might expect, are normally over a longer time frame, more strategic in nature and can influence issues such as design. Operational decisions by contrast focus on day-to-day issues associated with managing effectively and efficiently the product flow within the strategically planned supply chain. In addition, the authors suggest that there are four important decision areas in the management of a supply chain. They are location, production, inventory and distribution. Within each important decision area there are both strategic and operational aspects.

Further, they suggest that the decisions on the geographical location of production facilities, warehouses containing stock, and sourcing points are a natural first step when creating a supply chain. Such decisions involve the commitment of resources to a strategic plan. When the size, location and number of these are determined, the possible paths by which the product flows to the final customer can be identified. Such decisions, primarily strategic, represent the basic strategy by which customer markets are assessed in terms of revenue, cost and the service level. Reaching such decisions in a manufacturing supply chain requires the optimisation of costs, tax implications, duties, local content, distribution costs and production issues. For a good analysis of these issues see Arntzen *et al.* (1995).

Fearne (1996) discusses the issue of strategic alliances and the issue of supply chain management mainly from the perspective of the food industry. He draws the conclusion that despite a highly segmented market, in which innovation holds the key, the transient nature of consumer tastes and preferences means that manufacturers and retailers need to be able to respond rapidly if opportunities are to be fully exploited. He suggests that whilst mergers and acquisitions are likely to play an important part in the restructuring of the European food industry, strategic alliances provide a cheaper, more flexible and less risky means of maintaining growth throughout the food chain sector. This experience offers some interesting pointers to the construction industry.

11.3   Background and overview

Since the early 1990s partnering, lean construction and supply chain management have been increasingly advocated as important ways of improving the performance of the UK construction industry, both for clients and for the different members of the project team. These methods have been seen as an important way of dealing with the inherent problems of an industry still widely seen by many as a 'design to order' industry, craft-based and where each project starts with a 'clean sheet of paper'. Issues such as fragmentation, poor communication and a lack of integration, collaboration and trust remain to be resolved. Partnering, supply chain management and lean construction have been promoted as techniques that could be easily adopted from 'design to manufacture' industries such as the car industry and from retailing. It has been argued that these may be adapted to give considerable benefits to all partners involved, for example terms such as 'win–win' have been widely used by advocates. In addition, they have been a response by both the US and the UK to emulate levels of performance being achieved in Japan's construction industry. It will be seen that there are many definitions of all three of these techniques. In this chapter the term 'partnering' is used to refer generically to all such collaborative approaches (see Bresnen & Marshall, 2000). For a definition of lean thinking, in particular as it applies to construction, see the debate between Green and Howell (Green, 1999a; Howell & Ballard, 1999).

Since World War II a catalogue of reports have bemoaned the then current levels of performance of the UK construction industry and/or advocated change (e.g. Emmerson, 1962; Bowley, 1963; Banwell, 1964; Higgins & Jessop, 1965; Bishop, 1972; NEDO, 1975, 1988; Latham, 1994; Egan, 1998). Several (e.g. Flanagan & Norman, 1985) identified some improvements that had been made but concluded that these were patchy. Consistent themes can be discerned through the reports, for example fragmentation, short-termism, lack of trust and a lack of collaboration within the client/design/construction team. In addition, all have identified a lack of serious and sustained commitment to education, training, safety and research and in particular the low levels of commitment to serious skills development. These shortfalls were leading to consistently low levels of performance in areas such as cost, time, quality, running costs and fitness for the end-user.

11.4   Supply chain management and construction

It is argued that the rational decision-making of manufacturing cannot be emulated by construction unless the supply chain is strongly managed by the client. A definition adapted from the Construction Best Practice fact sheet on supply chain management (1998):

'Supply Chain Management (*in construction*) is a way of working in a structured, organised and collaborative manner, shared by all participants in a supply chain. Each company is a link in a chain of activities, adding value at each stage, designed to satisfy end-customer demand in a win/win scenario. The activities are all those associated with moving goods from the raw materials stage through to acceptance of the product or service by the end-customer. The process also embraces all the information systems necessary to support and monitor these activities.'

adapted from Construction Best Practice Programme (1998)

The Construction Best Practice Programme (CBPP) suggests that in UK industry generally, awareness of the strategic importance of supply chain management is high, but the implementation is low. However, there are significant differences between industry sectors. In particular, the construction industry does not feature among the more advanced sectors and small and medium enterprises in all sectors are lagging behind.

The supply chain encompasses all those activities associated with processing from raw materials to completion of the end-product for the client or customer. This includes procurement, production scheduling, order processing, inventory management, transport, storage and customer service and all the necessary supporting information systems. The supply chain includes, in construction, subcontractors and suppliers as well as the processes within the business itself.

'Construction businesses are beginning to realize that their success is increasingly dependent on the organizations they supply to and buy from, and that for continued success they need to cooperate and collaborate across customer/supplier interfaces.'

(CBPP 1998)

The primary purpose of supply chain management is to ensure reliability of quality and delivery time for a given product for an agreed price and at the same time to enable innovation and continuous measurable improvement in that product. Christopher (1992) suggests that the concept of supply chain management is little more than an extension of logistics and logistics management. Logistics management is an approach concerned with optimising flows within an organisation. Supply chain management recognises the greater flexibility and responsiveness required by increasingly sophisticated marketplaces and incorporates outsourcing and subcontracting.

Supply chain management strategies as they are adopted in the manufacturing industries assume an ongoing process where suppliers and customers frequently transact for the same, or similar, product. They are seen as key to maintaining quality and facilitating innovation and measurable improvement. To a large extent they depend for their effectiveness upon building a team culture for the particular benefit of the client and the mutual benefit of all members of the team. Supply chain management is seen as a vital part of the manufacturing processes and is rooted in operations management theory.

To use the term 'supply chain management' in the context of the current UK construction industry suggests that it is possible to adopt effectively those practices which have proved to be successful in the manufacturing sector without adapting them to reflect the particular nature of the industry and its culture. With the possible exception of housing, the context in the construction industry is that transactions are neither ongoing nor frequent and projects are usually unique and one-off in character. There is no production line. Most projects are procured by inexperienced purchasers and constructed by numerous specialists who have little or no contact with that purchaser. The business cultures of manufacturing and construction are also quite different. Consequently, to use the term 'supply chain' is at least curious and probably simply inaccurate if it is used in the manufacturing context. Nonetheless, the principles involved may well be highly advantageous to the construction industry in serving its clients.

Most construction projects are procured through a method by which a defined project forms the focus for a building process carried out by a contractor who (usually) obtains the work by bidding the lowest price for carrying out the project. The appointed contractor will outsource or subcontract the majority of the work to a number of relatively small subcontractors who will usually win the work on exactly the same lowest-price basis. The number of subcontractors will vary with the complexity and nature of the project. These contractors and subcontractors will focus upon meeting their contractual requirements for the lowest possible cost. There is likely to be only low commitment at best to the client's primary objectives or to any perceived project team. A parallel but separate supply chain managed by the client or client's project manager will in most cases include the procurement of the financial resources to support the project and the procurement of the design process itself.

As a consequence, the traditional supply chains which need to be managed in a construction project are complex and temporary, involving participants who do not contribute other than to complete their small, often isolated, part of a one-off project. A team culture focused on the particular needs of the client or the project rarely exists. Different approaches to managing such a supply chain are therefore required if the potential benefits are to be achieved. The culture within which supply chain management can be developed in the main does not exist in traditional methods but can be created if the value of such change can be shown to be significant.

Since the mid-1990s joint government and construction industry initiatives (Latham, 1994; Egan, 1998) have encouraged construction clients to adopt different strategies to procure projects. Whilst regular, experienced and informed construction clients have begun to adopt different strategies, there is little evidence to show that the numerically dominant inexperienced irregular purchasers have done so. By far the dominant strategy adopted is design–bid–build with the lowest bidder winning the work.

Adopting 'low bid win' strategies results in a number of well-documented and inevitable outcomes, particularly where the design is already established. They are:

- Production processes that are geared to lowest cost rather than to 'right first time' or to 'best value'.
- Bidding processes that encourage a culture where suppliers will agree to almost any parameter to get the work, later to strive to achieve a cheaper solution or a higher price.
- Inability and unwillingness to cooperate in specialist design, innovation or collaborative problem-solving.

Many of the weaknesses associated with such an approach include late completion, budget overspend and product defects which can usually be attributed to the rigid adoption of the 'low bid' strategy. Support for this view has come from the second survey of client satisfaction carried out by the Construction Clients' Forum (CCF, 2000). Alternatives do exist but require cultural change within the construction sector and its professions and critically in relation to the advice offered, particularly to inexperienced clients.

There is a tendency for construction professionals and in particular contractors to focus on cost and it is this focus which underpins and dominates the strategies adopted in managing the existing supply chain. Whilst cost is not irrelevant, most clients usually focus upon the value of their project in terms of the business case for it. It is this value or 'worth' which will form the key success factor for the project. The 'worth' may relate to the performance of the new facility in terms of its function or its worth as an asset. It may have a 'worth' in the marketplace at a particular time or over many years. There is a tendency for the long-term objectives of project worth or value to become refocused on the short-term objectives of cost and time once the project or strategic brief has been established (for a discussion on this see Ashworth & Hogg, 2000).

If improvements in performance are to be achieved, delivering a project which meets the primary criteria established by the business case will be most important. It is when these primary criteria are translated into key targets including cost limit and time that the first steps to sub-optimal performance can be taken.

Most construction clients will hope to achieve a high-quality project, cheaply in no time at all. The inexperienced client may believe or be convinced that this is possible. Analyses of industry performance show it to be unlikely. These key objectives need prioritising, so that the client understands that overemphasis upon one objective may put others at risk – speed may cost money and low price may result in poor quality. The client's project manager insisting that the client establishes a clear prioritisation of objectives may be a considerable step forward in achieving a context within which supply management can be achieved. Subsequent to the establishment of prioritised project objectives, sensible and achievable parameters for cost and time should be established and design developed in line with these parameters.

Cost estimates and time planning are very difficult for projects that are one-off. Such estimates of time and cost must be based on previous experience of similar projects. When these estimates are established, the growing team of consultants involved with the projects are regularly in an optimistic mood, keen for the project to progress and for their involvement in it so as to secure their own company's fee income/turnover targets.

Examination of construction industry performance suggests that long-term 'worth' can only be achieved by establishing parameters that are realistic and achievable. Once this has been successfully achieved, supply management techniques can be adopted to obtain the best out-turn.

One key aspect of traditional supply management in the manufacturing sector is the involvement of production specialists in the design process. Most of the work involved in a construction project is carried out by specialists, usually employed as subcontractors to a main contractor. It is at this design stage that the involvement of specialists can be most beneficial – when design is carried out and cost and time parameters set. Because of the contractual link between these specialists and the main contractor they are usually precluded from offering advice to the client's team at this early stage. Although it is possible through a process of nomination to have some specialist involvement, there is a limit to the extent of nomination that can be realistically achieved without exposing a client unreasonably to increased risk. In a traditional design–bid–build construction project, supply management is carried out only by the main contractor through a subcontracting process. This is usually even more price-focused than the initial contractor's tender, as the contractor seeks to negotiate an improved financial position for himself with the subcontractors he intends to employ. Consequently, the dominant traditional approach to construction procurement does not involve anything other than management of supply by contract specification, for a competitive price.

If the benefits of specialist involvement in the design process, including increased innovation and collaboration, are to be achieved through supply management, traditional processes and attitudes will have to be abandoned with a new culture focused on the primary objectives and 'worth' of the project to the client.

Strategies that may be adopted to facilitate supply management throughout the design and construction phases of a project include strategies where the client can participate or strategies facilitating collaboration. The single most suitable procurement strategy, which will enable client participation in managing the supply chain, is construction management. This is where the client or his project manager employs trade contractors direct. This approach will, however, require constant participation by the client who may not wish or be technically competent enough to have that level of involvement. In such a case the employment of a client's project manager is vital. Experienced clients who can lever advantage through repeat business so as to offset the extra fees incurred therefore mainly adopt this strategy.

11.5   Construction – supply chain management and partnering

Partnering or partnership sourcing is normally a major feature associated with successful supply chain management. The benefits to the construction industry include having a better understanding of client needs and of suppliers' capabilities. This leads to greater predictability and potential reduction in time for completion and cost of production, with a parallel improvement in quality and reduction in the amount of defects.

Many cases for partnering can be found in the literature, but more recently some concerns have also begun to emerge. In addition there are many definitions of partnering, perhaps because of the very different 'world views' of the various authors. The concepts of 'project partnering' and 'strategic partnering' can be identified. There is broad agreement in the literature about the overall philosophy of partnering. There are, however, widely divergent views on a number of its features, leading to the view that it is like many other management concepts in being imprecise, inclusive and subject to continuous redefinition. It is difficult at times to distinguish between partnering as a distinctive practice and partnering as managerial rhetoric.

From the literature, the conclusion is drawn that there is wide agreement over the benefits of partnering as a project procurement strategy for the UK construction industry. Many see it as far more than just a procurement strategy, rather a fundamentally new way of doing business. However, this is somewhat weakened by a lack, so far at least, of rigorous, verifiable evidence to support claims that are made. This is despite the carefully argued need for such evidence in the literature. It is clear that in-company evidence is available, as a company will claim to know what represents value for its business and most have remained enthusiastic after a number of projects. Such evidence, if independently audited, would add serious weight to claims of benefit. In addition, some of the benefits and concerns that have been identified have clear implications for public sector clients who are using a partnering procurement strategy (see Fisher & Green, 2000).

Cultural changes needed if partnering is to deliver benefits

One consistent theme drawn from the long list of government and industry reports discussed above is that conflict is very much the industry norm. Non-cooperation, based on fundamentally different worldviews and interests between clients, designers, engineers, cost consultants and contractors, is a major characteristic of the construction industry. Often it is the issue of commercial pressures, and in particular traditionally low profit margins, that force partners to act in traditional, adversarial and exploitive ways. Economic conditions that encourage collaboration are an important factor in forcing contractors to accept change. For example, a buyers' market, where powerful clients or client groups force changes on contractors in terms of ways of working or risk acceptance, has been observed in the oil and gas sector (Bresnen & Marshall, 2000). But this is hardly the spirit of partnering. The opposite could also be true. Other issues are raised in the literature such as the building of trust, project team building, the need for top-level commitment, the role of the individual and the need for open and flexible communication (Barlow *et al.*, 1997).

Given that trust is identified in the literature as one of the cornerstones of successful partnering, it is surprising that many of the advocates of partnering in the construction industry have notably ignored the wider management literature on trust between business organisations. The partnering literature assumes that trust within organisations is the same as trust between organisations, ignoring the broader institutional constraints that impinge upon relationships. Some advocates of partnering even make simplistic analogies with marriage and personal relationships. Discussions of partnering arrangements based on inter-organisational trust are founded on the assumption that organisations are unitary entities sheltered from environmental influences. Blois (1999) argues that trust can only be granted by individuals. The argument that partnering depends upon trust between organisations is therefore interpreted as short-hand for two sets of individuals each of whom is trusting the organisation of which others are members. Individuals would clearly be unwise to adopt blanket trust of organisations that are subject to short-term economic pressures. Such economic exigencies are always likely to take precedence over the sensitivities of middle managers. The likelihood of individuals trusting an organisation will be further significantly shaped by the organisation's reputation for trustworthiness (Kreps, 1996). Of particular importance is the way that an organisation has behaved when faced with unexpected contingencies. Some firms will protect their reputation by adhering to principle even when it is not in their short-term interest. Individuals are unlikely to trust organisations who, despite an overt commitment to partnering, revert to form when faced with an unexpected occurrence. Some of the major advocates of partnering possess reputations for fair practice that are currently being questioned. Claims to have suddenly 'seen the light' should be, and invariably are, treated with suspicion. If partnering is dependent upon trust, it must be recognised that this in turn is dependent upon behaviour over prolonged periods of time. The behaviour of organisations and individuals within the construction industry will not be changed by exhortations from the Construction Industry Board or others.

There is evidence that partnering can depart from the ideal if specific concerns dominate the thinking of the powerful members of the partnership, for example a drive for cost reduction, or determined attempts to push risk or cost reduction down the supply chain. Indeed, where the sole purpose of the partnering exercise is to improve performance on a continuous basis, low margins may result in the use of power to squeeze suppliers or subcontractors too hard, in which circumstances, as Bresnen and Marshall (2000) put it, 'there is the paradoxical danger that partnering could become a victim of its own success' (see also Bresnen, 1996; Green, 1999b). Green (1999b) examines some of the companies put forward as illustrations of good practice and expresses concern over the track records of a number of them, partnering with suppliers in other, non-construction areas of their business. He points out that several of those cited are currently under investigation by the UK Competition Commission.

There is evidence of an unwillingness to commit fully to close long-term partnerships because they inhibit open price competition and the use of alternative suppliers. Rasmussen and Shove (1996) put this point neatly when they suggest that there is no easy solution based on exhortations to act in ways that fly in the face of powerful economic imperatives and well-established traditions. Clearly, the benefits of partnering are going to be relative, as it will serve different interests for different parties. Bresnen and Marshall (2000) build on this by demonstrating from the literature on organisational and cultural change the complexity of the task facing those wishing to develop a suitable organisation and culture for partnering. It is likely that short-term changes can only be achieved as a result of the economic/power issues raised above. They go on to suggest that there is serious doubt in the literature about the possibility of manipulating or changing organisational cultures, as they are not simple variables like organisational structures or reward systems. This view contrasts with that of Carlisle (1991) and Axelrod (1984) who argue that cultural change is possible. A culture is the very essence of what a company or an industry is. Bresnen and Marshall (2000) further suggest that some current conditions do assist any pursuit of change in the construction industry; however, there are others that still pose major barriers.

Korczynski (2000) has compared the factors associated with a low-trust economy with those for a high-trust economy. A high trust economy is one that rates highly a trusting social and ethical framework within a long time horizon and in which a good reputation goes hand-in-hand with high status and repeat business. A low-trust economy is characterised by organisations which are economically opportunistic within a short time horizon, exist within an inefficient enforcement regime, and accept, as normal, a poor reputation, and a low likelihood of repeat business. The structural characteristics and ingrained practices of the construction industry would seem to accord much more closely with the factors associated with a low-trust economy. The industry is notoriously short term and narrowly rational. The market is further characterised by significant power imbalances throughout the supply chain. The possible weakening of professional and trade associations may be indicative of a construction industry that is moving even further towards a low-trust economy. The project-based nature of the industry makes the likelihood of repeat work relatively small beyond the industry's large clients.

Barlow *et al.* (1997) identify the fact that there is no one formula for setting up a partnering agreement, as the type of project, the degree of risk, preferences of clients and other specific factors will determine the form it should take. A contingency approach is suggested. Such factors will also determine contractual arrangements. However, in nearly all the projects that Barlow *et al.* (1997) observed, a standard contract was somewhere behind the innovative process. Bennett and Jayes (1995) deal with the various contractual questions, looking at issues for the public sector and methods of adjudication. In addition they look at the issues caused by the use of partnering in the light of current EU competition regulations. Several examples observed by Barlow *et al.* (1997) had a full dispute resolution procedure agreed, including a committee to deal with disputes. However, Barlow *et al.* did observe an increased level of trust developing between parties, with the contract seen more as a safety net. This attempt to improve trust was strongly advocated by Egan, who urged companies to work together on a basis of trust (Egan, 1998).

The majority of clients, being relatively small and inexperienced, depend heavily on their construction professionals who are often seen as barriers to, rather than initiators of, innovation. Additionally, Latham's interim report *Trust and Money* (1993) indicated that the construction industry lacked evidence of trust, so the requirement for trust is particularly difficult to achieve. The partnering philosophy encompasses the development of mutual objectives, an agreed method of dealing with disagreements and continuous measurable improvement. It depends upon the development of collaboration and trust between parties who have for many years existed in a climate of competitive commercial contracts and adversarialism. The parties must see that the commitment from the top is there and that there are real tangible advantages, such as an agreed profit level, shared benefits from innovation and a wish to avoid expensive and time-wasting litigation. This ethos is consistent with many of the philosophies successfully adopted in manufacturing industries where the processes and products are improved on a continuous basis (see Fisher & Green, 2000)

Supply chain management can thus be improved by the adoption of collaborative (part-partnering) strategies and techniques without significant variance to established procurement strategies. These techniques involve serious collaboration between the client's team and the construction team in improving estimates of cost and time and in enabling a level of constructor involvement in the design process.

11.6 Established procurement systems and supply chain management

There are a number of established procurement systems that will enable collaboration between the client, his/her consultants and constructors to the benefit of the supply chain management process, the extent of the collaboration being dependent upon the strategy adopted. Two-stage tendering based upon partly completed design enables a largely traditionalist competitive approach to contractor selection but facilitates some constructor involvement once the contractor has been pre-selected on the basis of prices against approximate quantities or similar criteria. Constructors can advise on the likelihood of the project meeting its desirable completion date and will be able to assist the design team in terms of the practicability of the detailing. This level of involvement may enable some improvement in the accuracy of time management and level of buildability but may be too late in the process for much to be gained in terms of value for money through design review. Nonetheless, more certainty in terms of management of construction time may be achieved and some benefits associated with establishing critical design activities may ensue.

On the other hand, the selection of procurement strategies such as management contracting facilitates a high level of constructor involvement enabling design buildability, value management and concept briefing and the benefit from early constructor involvement in terms of estimates of construction programme and cost. The adoption of these strategies is numerically very low, however, when compared with traditional and design-and-build strategies.

The adoption of any of these strategies facilitating collaboration can be quite a step for some organisations, usually because contractual risk shifts back to the client. However, by accepting a little more risk through the adoption of collaborative strategies, predictability may be increased and consequently overall risk reduced. It is in this environment that prime contracting has developed. Prime contractors understand better than others the logical progression of construction and how long construction operations take and how much they cost. Their involvement will therefore improve predictions and may improve the performance of the whole team in setting achievable targets, achieving better value for the client and improving feedback loops between constructors, subcontractors and designers. Similarly communication and feedback between the client and the prime contractor will enable trust to grow leading to, if not full partnering, then collaborative practice. A simple step in drawing together elements of a fragmented industry which have traditionally regarded each other with suspicion and mistrust provides an opportunity for a win-win solution.

Most construction work in the UK does not currently adopt supply chain management principles and therefore most construction clients do not gain any of the benefits such processes may offer. Whilst the minority of experienced, regular, big-spending clients have used their buying power to lever supply chain benefits through adopting partnering strategies, this is counter to the deep-rooted culture adopted in most other cases. Unless less-experienced clients are encouraged or advised to adopt more collaborative procurement strategies focused upon project worth rather than project cost, it is unlikely that the business climate in construction will change. To date, there is little evidence of the attitude of most clients or their professional advisers changing to adopt current best practice thinking.

11.7   A case study – supply chain management and the construction process

In 1997, UK Defence Estates (DE) within the Ministry of Defence combined with the Department of the Environment, Transport and the Regions (DETR), to set up and sponsor 'Building Down Barriers' (BDB). They wished to create a learning mechanism for establishing the working principles of 'supply chain integration' in construction. The first phase of the initiative, which is also sponsored by AMEC and Laing, was finished during 2000 (see Holti *et al.*, 2000).

Holti suggests that supply chain integration is the cornerstone of the BDB approach. Also that integration should be approached from two perspectives: (1) those who design and those who construct/deliver need to be brought together; and (2) the supply chain needs to be kept together over time, from project to project. A facet of the latter is that during a project the typically fragmented accountabilities of participants need to become one frame of reference. Also, the supply chain needs to be organised so that a building is designed to be economical to build, maintain and operate from the start, as well as meet client/user needs and other design requirements. The subsequent provision of the facility should focus on the aspects of quality that matter to a client and the users. In addition, that the full benefits of this integrated approach will come only from its repeated application by a supply team working together on successive projects. However, there is no need for successive projects to be for the same client, although as in strategic partnering, working for the same client does provide ideal conditions for project-on-project improvement (Fisher & Green, 2000). As Holti *et al.* say:

> '... supply chain integration and continuous improvement pose profound challenges for established patterns of responsibility in UK construction. In particular these concepts fly in the face of norms whereby consultant designers undertake design work that

is then priced by specialist subcontractors who have not been significantly involved in developing the design.'

Holti *et al.* (2000)

The BDB project was based on two key issues:

(1) A rigorous structured project process
(2) A collaborative model of leadership

BDB comprised a research/development team which devised the BDB process. The research team then facilitated and evaluated its use on two pilot projects. The projects were:

- The Aldershot Garrison Sports Centre which formed part of the army's centre of sport excellence at Aldershot. The £10.8 million project consists of a major swimming pool, two sports halls and facilities for a large range of sports.
- The Wattisham Physical and Recreational Training Centre which is at Wattisham Garrison in Suffolk. The £4.2 million project contains a swimming pool, a sports hall and squash courts. It will be used also for helicopter crew survival training.

Both pilot projects were completed during 2000.

As a result of the theoretical work and the fieldwork testing, a number of generic principles were identified. These principles are key to the harnessing of both the design and the delivery management leadership of the integrated supply chain. The adoption of these principles will necessitate major culture changes, in particular in the areas of attitude and behaviour by all groups and individuals party to a project. Holti *et al.* (2000) identified seven principles, but it is important to note that in committing to the first principle all parties must also be committed to the other six 'as a mutually reinforcing set'. The seven principles are:

(1) *Compete through superior underlying value* – that is, better value for the client through both a lower price and better quality in terms of what is best for the client. This is the key principle and all others relate to and add to this one.
(2) *Define client values* – that is, rejecting as too narrow the single criterion used to assess most construction decisions in the UK, namely the capital price. Rather it attempts to look at whether the facility really meets the needs of its users and whether it represents good value for money in terms of how much it costs to run and maintain. BDB attempts to set the standard for gauging the value of what is being delivered in two ways.

Firstly, functional requirements are made as clear as possible and secondly, the cost of providing these functions are measured, that is, defining client need in terms of output and designing for through-life cost performance.

(3) *Establish supplier relationships* – according to the BDB team, typically the products and services provided by the companies in the supply chain account for in excess of 80% of the total cost of a construction project. The way in which they are procured and delivered can critically influence the outcome of the project. BDB identified a key issue for prime contractors, namely that of demonstrating commitment to forming long-term relationships with companies which will be the major suppliers of products and services to the kinds of construction project they see as making up their core business. Once achieved, this can both drive up quality and drive down capital cost and through-life costs for clients, whilst at the same time achieving increased profitability for the supply chain.

(4) *Integrate project activities* – that is how the prime contractor manages the construction project. The BDB team devised a technique to identify which suppliers are the critical long-term partners through whom the effective management of suppliers on a project can be best achieved.

(5) *Manage costs collaboratively* – this approach is called target costing, where the supplier works backwards from the client's functional requirements and the maximum market price for the item. Project teams design a product that meets the required level of quality and functionality but also provides a required level of profit at that target price. Costs are thus managed before they are generated, to protect margins and provide the security to look at underlying cost.

(6) *Develop continuous improvement* – this is the vehicle for the achievement of longer-term performance improvement, that is, in terms of what is delivered to the client and of profitability of the whole supply chain. It is the philosophy that is behind the concept of total quality management. All of this in practice requires greater attention to planning how to do things in advance of construction so as to anticipate and avoid problems occurring on site.

(7) *Mobilise and develop people* – when devising their process the BDB team were keenly aware that developing long-term supply relationships, in the context of both continuous improvement and at the same time carrying out individual projects using the new BDB approach, would be extremely challenging for those in the UK construction industry. This is primarily because they were managing a comprehensive

programme of organisational change within their parent company based on the principle of collaboration with key suppliers
not in conflict. Such a new approach would involve a massive
culture change. The BDB team identified four key drivers of
change both within prime contractor organisations and the full
supply chain. They are: (1) visible systematic commitment from
the top; (2) facilitation for project teams; (3) training in new
skills; and economic incentives.

The effective implementation of these seven generic principles
requires something extra, a new project process, but one developed
in a structured way that is capable of integrating the whole supply
chain.

11.8 Conclusion

There are issues associated with the benefits and risks of engaging
supply chain management that have clear implications for those
charged with protecting the client and in particular the taxpayer.
Better value for money is one of the much-trumpeted benefits of
supply chain management. However, one concern is that as the
supply chain process becomes more established and effective and
the benefits become evident, so too does the domination of the
process by those who control and manage the supply chain. This is
already becoming all too apparent in the food industry where the
control of the major supermarkets and their logistics operation over
food producers and suppliers is causing concern even at government levels.

There is a culture in the public sector that is based on the idea that
one-off competitive tendering is the safest way to get value from
public money. However, from a public spending viewpoint, there is
merit in allowing the marketplace, in terms of partnering and supply chain management, to work its course and disprove this notion.
If this policy is to be followed, then those charged with public
spending need to be equipped with both appropriate tools to
identify and sanctions to protect the taxpayer against anti-competitive behaviour, such as can result from integrated supply chains.

However, the growth in PFI projects offers serious opportunities
and challenges to the project procurer in the public sector. The use of
partnering and supply chain management as a procurement strategy or as a new, fundamentally different way of doing business, is
widely claimed to be a serious option for achieving better value for
money for the taxpayer, provided the concerns outlined above are
properly dealt with. The public sector as customer has an important

role in influencing the growth of such networks. Recognition of this is in contrast to the prevailing practice in the public sector of using its power to foster competition *within* project teams, with often serious consequences for the taxpayer.

This chapter suggests that there is evidence to demonstrate that there are measurable benefits in using supply chain management (and partnering) as a project procurement strategy within the UK construction industry. Some see all this as much more than a procurement strategy and suggest the term 'paradigm shift'. Independently verified evidence of measurable benefits is currently lacking in sufficient quantity and breadth to be convincing; simply, it is just too early. However, it is important for any measure to be independently verified, so as to avoid charges of 'corporate marketing propaganda' or 'current fashion'. Clearly there are dangers of overplaying both the benefits and the ease of implementation and of attempting to provide a standard supply chain management solution to all project situations. There is also a danger of a counter-attack against the supply chain concept by traditionalists, who have their own reasons for supporting the status quo. But as Green (1999b) points out, with major clients pooling their buying power and demanding improvements, it is not surprising that there are longer-term concerns over both partnering and supply chain management in general, as witnessed in the allegations against major food retailers recently under investigation. Simply put, if powerful organisations grab too much of any benefit, it is against the spirit of trust central to both supply chain management and partnering.

In addition, there is an urgent need for those charged with public spending to deal with the issue of barriers to entry. They must also be equipped with appropriate tools to identify and sanctions to protect the taxpayer against anti-competitive practices when they occur. Exactly when do integrated supply chains, or barriers to entry become anti-competitive? This is an issue that must be balanced, against the fact that 'cut-throat' competition also poses just as real a threat, with quite disastrous long-term consequences for all.

References

Arntzen, B.C., Brown, G.G., Harrison, T.P. & Trafton, L. (1995) Global Supply Chain Management at DEC. *Interfaces*, January.

Ashworth, A. & Hogg, K. (2000) *Added Value in Design and Construction*. Longman.

Axelrod, R. (1984) *The Evolution of Co-operation*. Basic Books.

Banwell, H. (1964) *The Placing and Management of Contracts for Building and Civil Engineering Works*. HMSO.

Barlow, J., Cohen, M., Jashapara, A. & Simpson, Y. (1997) *Towards Positive Partnering*. The Policy Press Bristol.

Batchelor, C. (1999) Logistics aspires to worldly wisdom. *Financial Times*, 17 October, p. 15.

Bennett, J. & Jayes, S.L. (1995) *Trusting the Team*. Centre for Strategic Studies in Construction, University of Reading.

Bishop, D. (1972) Building technology in the 1980s. *Philosophical Transactions of the Royal Society*, **A272**, 533–63.

Blois, K.J. (1999) Trust in business to business relationships: an evaluation of its status. *Journal of Management Studies*, **36**(2), 197–215.

Bowley, M. (1963) *The British Building Industry*. Oxford University Press.

Bresnen, M. (1996) An organisational perspective on changing buyer–supplier relations: a critical review of the evidence. *Organisation*, **3**(1), 121–46.

Bresnen, M. & Marshall, N. (2000) Partnering in construction: a critical review of issues, problems and dilemmas. *Construction Management and Economics*, **18**, 229–37.

Carlisle, J. (1991) *Co-operation Works – But It Is Hard Work*. John Carlisle Partnership.

Christopher, M. (1992) *Logistics and Supply Chain Management – Strategies for Reducing Cost and Improving Services*. Pitman Publishing.

Construction Best Practice Programme (1998) http://www.cbpp.org.uk

Construction Clients' Forum (2000) *Client Satisfaction Survey*. CCF.

Cook, P. (1999a) *Building*, 5 November 1999, The Builder Group.

Cooper, M.C. & Ellram, L.M. (1993) Characteristics of supply chain management and the implications for purchasing and logistics strategy. *The International Journal of Logistics Management*, **4**(2), 13–24.

Egan, Sir John (1998) *Rethinking Construction* (the Egan Report). Department of the Environment, Transport and the Regions, HMSO.

Emmerson, H. (1962) *Survey of Problems Before the Construction Industries*. HMSO.

Fearne, A. (1996) Strategic alliances and supply chain management. Paper given at the *2nd International Conference on Chain Management in Food Business*, ISBN 9067544507.

Fisher, N. & Green, S. (2000) Partnering and the UK construction industry the first ten years – a review of the literature. In: *Modernising Construction*, HMSO.

Flanagan, R. & Norman, G. (1985) *A Fresh Look at the UK and US Construction Industries*. Building Employers Confederation.

Ganeshan, R. & Harrison, T.P. (2000) *An Introduction to Supply Chain Management*. Pennsylvania State University.

Green, S.D. (1999a) The dark side of lean construction: exploitation and ideology. Paper given at *ICLG-7*, University of California, Berkley, pp. 21–32.

Green, S.D. (1999b) Partnering: the propaganda of corporatism? *Journal of Construction Procurement*, **5**(2), 177–86.

Higgins, J. & Jessop, N. (1965) *Communications in the Building Industry*. Tavistock.

Holti, R., Nicolini, D. & Smalley, M. (2000) The Handbook of Supply Chain Management. CIRIA, ISBN 0 96017 546 4.

Howell, G.A. & Ballard, G. (1999) Bringing light to the dark side of lean construction. Paper given at *ICLG-7*, University of California, Berkley, pp. 33–38.

Korczynski, M. (2000) The political economy of trust. *Journal of Management Studies*, **37**(1), 1–21.

Kreps, D.M. (1996) Corporate culture and economic theory. In: *Firms, Organisations and Contracts* (eds P.J. Buckley & J. Mitchie), pp. 221–75.

Latham, M. (1993) *Trust and Money*. HMSO.

Latham, Sir Michael (1994) *Constructing the Team*. (the Latham Report). HMSO.

National Economic Development Office (1975) *The Professions and Their Role in Improving the Performance of the Construction Industry*. HMSO.

National Economic Development Office (1988) *Faster Building for Commerce*. HMSO.

Rasmussen, M. & Shove, E. (1996) *Cultures of Innovation*. Working paper, Centre for the Study of Environmental Change, Lancaster University.

12 The Management of a Project

Margaret Greenwood

12.1 Introduction

The constraints within which the project must be managed are set by the definition of the project (the expectations and the boundaries), the project brief and an outline of the time, cost and performance constraints established in the business plan. In addition project management addresses the resources and events that must be monitored and controlled and the quality that must be assured. Particular attention is also given to the motivation and management of people that are and will be involved in the project.

This chapter outlines the role of the client-appointed project manager. The project manager, whether the client, an employee of the client or an external consultant to the client organisation, must be fully informed about the project, the stakeholders, the constraints and the resources and must be given appropriate and sufficient authority and accountability to mobilise resources and make decisions regarding the project. The project manager will carry out and/or direct necessary functions, systems and controls to achieve project completion. The client's role is one of formal approval of and support for the project manager's actions, ensuring that the necessary resources, organisational structure and contractual arrangements are in place for a successful project outcome.

Whilst this chapter cannot be a comprehensive manual on project management, its intention is to outline the role of the project's manager. It highlights project planning and implementation policy, setting up the project and mobilising resources and specific management requirements. Attention is also drawn to the *Code of Practice for Project Management* published by the Chartered Institute of Building (1996).

12.2   The project plan and the implementation policy

Three major tasks are the first concern of the project manager:

(1) To develop the project plan within the constraints and resources specified by the client
(2) To identify the project team who will carry out the plan to successful fruition
(3) To initiate and implement the procurement strategy

Developing the project plan

Earlier chapters have referred to the importance of the briefing process, project definition and procurement strategy. Developing the project execution plan from this information is the key primary role of the project manager. As a first stage it is advisable to identify all stakeholders (those parties interested in the project's outcomes), interpreting their strengths, interests and influence in the project outcomes. These may be positive or negative. It is also crucial to identify and disclose all constraints. The client must state:

• The total funds at the client's disposal for this project, their availability, the real (if any) limits to overspend, and any risk to the availability of the funds.
• The required completion date, any leeway (positive or negative), and significant points (gateways or checkpoints) at which progress should be reviewed.
• The required quality or performance of the final product.
• The client's attitude to the risk inherent in the project process

An approximation of the resources requirement is also necessary at this early stage as this may impact on the project specification and plan. It is unlikely that the detail, of man hours for example, will be available at this stage (detailed design work may yet have to be carried out) but an estimate of 1000 hours as opposed to 10 hours may well impinge on the constraints.

Identifying the project team

The project manager will need to be supported through the project by a team – a group of people with particular skills, capability, experience and knowledge who are specifically appointed to carry out the project plan to successful completion. The team will assist in monitoring, controlling and generally managing the project but will not necessarily carry out the construction work. Levels of authority and accountability for team members must be defined. At the earliest stage it may be advisable for the project manager to appoint certain 'experts' to supplement the project management skills e.g. procurement specialist, specification writer, planner/programmer.

Implementing the procurement strategy

This is a key and early role and warrants dedicated thought and review; it is referred to in Chapter 10. A policy must be developed and implemented to ensure appropriate procurement in light of the specification and plan (see above). It must be clearly communicated to all the key players. Special attention should be given to maintaining this policy throughout all stages of the project. This will ensure the client has a greater chance of obtaining the right project at the right time at the right cost with the minimum of conflict.

12.3   Setting up the project and mobilising resources

Financial resources

Normally the funding function is regarded as a matter for the client. A specific individual, company or manager may be appointed by the client to take over this role in which case boundaries of accountability and responsibility must be acknowledged. Funding is a vital prerequisite to the project. By the time the procurement strategy is set, sufficient funding must be planned to be available at appropriate times in the pre-construction and construction process. The means of funds availability must also be planned. Funding requirements are a consequence of contractual agreements, whether they be agreements to purchase land, to design, to construct or to project manage, etc. Once the project plan is in place, it is possible to take account of each contractual agreement and to plan expenditure in the form of a cash flow. In this context the project manager, perhaps with support from the project cost consultant, should take control to ensure that funding arrangements are appropriate.

Human resources

The human resource element required for the completion of the project may be viewed as three distinct groups: the client team, the project team and the construction team. The client must be prepared to allocate appropriate in-house personnel to the project, or to appoint consultants for this purpose, to uphold his/her brief within the constraints. The client may, as in most current construction projects, appoint consultants to design and cost a project, but the client should have his/her own staff member to provide liaison and a focus for decisions. If this is not possible it is desirable, in the interests of expediting the project, to explicitly hand these duties over to the appointed project manager.

The project team may be part of or external to the client and/or the project manager's organisation. The team should be appointed by and be accountable to the project manager, on behalf of the client. Each team member will have a specific task and/or role and their level of authority and accountability must be defined. Membership of the project team may vary throughout the project as the requirements for each stage differ.

A main contractor, construction manager or an equivalent will normally lead the construction team. It is important that all direction and control of this group be made through this designated person. If this is not carried out and controlled there is likely to be mis-information, confusion and conflict that can potentially lead to project failure. Many projects fail because of a lack of sufficient attention to people – that is, both individuals and groups. Under-standing and communicating agendas and requirements help to build a level of trust and openness necessary to ensure the client receives a successful and complete project.

Physical resources

Other resources such as land, buildings, plant, machinery and materials are needed for project completion. At an early stage it is necessary to identify who is responsible for making these available and when and where. Some often-encountered problems are high-lighted here.

(1) Where physical resources such as land, plant and machinery have to be provided by the client, the client must ensure their availability in a suitable form at the right time; failure to do so could be costly and will impact on success.

(2) Should the client require alternative accommodation, or where specific arrangements for decanting an office and staff are necessary, this is normally a matter for the client and/or his organisation. Alternatively, these requirements may be part of the project but should be costed and contracted for accordingly.

(3) Design resources must also be appropriately selected and in place at the right time. Normally this is the responsibility of the client but may be part of the project manager's brief. The criteria for the selection of the design team should include factors of capability, competence, staff and cost. Value for

money, rather than cost, should be the major influence in selection.

(4) Construction resources will be contracted by the client or on behalf of the client by the project manager, and must be appropriately selected and in place at the right time. The criteria for the selection of constructors are broadly the same as for the selection of consultants but the process tends to be much more complex. Selection alternatives are:

- Through competition, based upon a fixed design and specification or, in the case of design-and-build options, based upon competition for design solutions.
- By negotiation with one or two constructors.
- A two-stage process involving one or both of the above processes.
- A partnering philosophy designed to enhance the commitment of all members of the supply chain towards benefiting the client's objectives and business case.

(5) Suitable documentation for the selection process must be prepared and the client is advised to seek the advice of the project manager. He/she may also be required to manage the process.

Note that a number of Codes of Procedure detailing selection processes are published by the Construction Industry Board (1997a–g). These form the basis of best practice in constructor and consultant selection.

Organisational structure

A 'temporary organisation' is created for the life of the project to enable effective communication and decision-making, management of client input, coordination of functional and administrative needs and the resolution of conflicts. It also acts as the formal point of contact for the project. This organisation may be headed by a key member of the client organisation or more usually by the project manager. This designated person must have sole authority to communicate decisions to the project design and/or construction team. If more than one individual is empowered to instruct, or to require changes, confusion will occur.

Building users, specialists, facilities managers, maintenance staff, finance and account personnel, legal advisers and security personnel can all have input to the project through this temporary organisation by invitation or by right. As appropriate, it can meet with designers or constructors to ensure effective communication. All decisions, however, should be exclusively reserved for the client's executive or the project manager.

Contractual arrangements

Having selected an appropriate procurement strategy it must be implemented. A range of contractual agreements are available, many of which are based upon standard forms of agreement already in common use in the construction industry. The project manager will normally give advice upon the implication and selection of the appropriate form of contract.

Managing the project – an overview

Once decisions have been made about the project plan, resources, the organisational structure and contractual arrangements, the project manager is responsible for implementing the procurement process and managing the project to its successful completion. These systems enable the project manager to inform the client and other stakeholders of the current position at any point through:

- Financial systems that will ensure payments are made in accordance with contract agreements.
- Decision systems that will ensure decisions are communicated at the appropriate time and with the appropriate authority.
- Design change systems that will implement and monitor change.
- Cost and time monitoring systems that will chart real progress against the plan.

Controlling these systems to ensure that they are appropriately applied and not abused is a matter for the project manager. The diversity of these systems may mean delegating elements of the process to members of the project team (if appointed) or perhaps ensuring that existing client systems are specifically adapted. Methods of control ensure that the costs remain within the budget, the pace stays within the programme, the quality or performance is within the standards set and all the systems are properly employed.

The project manager or an appointed cost consultant will achieve cost control by ensuring appropriate pre-construction cost planning and an adequate system of regular cost reporting. Slippage will be immediately visible and hence easily manageable.

Pace control can be achieved by monitoring progress against the project programme or time plan. Progress should be assessed against imposed gates, checkpoints or contract completion dates. Again, slippage can be immediately seen, communicated and action directed.

Quality and/or performance can be controlled through the provision of a sound contract that includes a detailed specification and by appointing constructors and consultants who have quality assurance systems in place. It may also be appropriate to appoint a dedicated quality control person, such as a clerk of works. His/her role would be to inspect the construction work during the progress of the project, observing and measuring quality and thereby ensuring ongoing work is satisfactory and matches the specification.

12.5   Specific aspects of managing the project

Time control overview

The overall programme, or time plan, of activities that constitute the project must be developed at an early stage in the procurement cycle. It is common for IT-based programming systems to be adopted in carrying out this process and there are a number of effective and reliable systems of this type currently available.

The programme forms the framework within which the designers and the constructors must start and complete their activities. It also serves as the framework within which other key stages, such as land purchase, funding and planning permission should be completed. This programme of activities should be feasible and realistic. Insufficient design time will have an effect upon client risks, as will insufficient time allowed for construction. The planner must be aware of those tasks which are critical to completion on time and those tasks where some flexibility may be available. Time for approval processes should also be included as specific activities; this is invariably vital, as the need to obtain approval is likely to be a prerequisite to proceeding to the next stage of the project and further work.

The process of time control is in many ways analogous to that of cost control. Thus, a time control system should embrace:

- A time budget – the overall project duration as fixed either by specific constraints or by the selected procurement strategy; this period becomes a key parameter for management of the project.
- A time plan – the division of total time into interlinked time allowances for readily identifiable activities with definable start and finish points.
- The overall project programme – linking the above.
- A time checking system – the actual time spent on each activity must be monitored closely against its allowance; divergence must be reported as soon as it is identified.

If an activity exceeds its time allowance there are essentially only two forms of corrective action available:

(1) The resequencing of later activities. This may involve abandoning low-priority restraints and/or it may require phased transitions from earlier activities to later activities which logically follow them.
(2) Shortening the time allowance for future activities. This can usually be achieved by increasing specific relevant resources.

If neither is done, the overall time budget is likely to be exceeded and the project will finish late.

It should be noted that there is a risk that certain unavoidable events (e.g. flooding) may delay the overall completion of a project. Such possible events should be identified and contingency provision should be made for these effects to be taken into account as soon as possible, taking cognisance of contractual conditions. Other measures may also be considered, should the client so wish, in mitigation of serious delays and associated costs.

The fundamental relationship between time, quality and cost should recognise that:

- Any extension of the overall time-scale for a project always generates additional costs to constructor and/or client. Every project contains time-related costs whether these are openly stated or not. The decision as to who carries such additional costs depends on the detailed contractual arrangements between the parties; it is likely that some of them will be borne by the client.
- Making up for lost time by resequencing later activities may be achievable but is often only at the risk of compromising quality or cost control.

Time control is as important during the design stages of the project as the construction stage. Designers must work to a series of deadlines (gates) at which different elements of the design should be agreed (i.e. frozen) in order that costs and the overall programme be kept under control.

At a very early stage in the project it is advisable for the client and/or the project manager to consider developing a time contingency (or reserve) with strict procedures for its allocation to specific events. This concept is essential on projects which are subject to external time constraints, for example where a building has to be available for occupation before the lease on another building expires.

Design overview

Whilst the responsibility for achieving a successful design solution to the client's requirements lies chiefly with the designer, responsibility also rests with the client organisation for ensuring that client needs are met and for the impact of the development on the local environment. The blame for a poor building is just as likely to be ascribed to the client as to the designer. Equally, a well-designed building reflects credit on those who commissioned it as well as on the designers who conceived it.

Design is an important factor in ensuring good working conditions for staff and convenience for members of the public who need to visit the building. A well-designed building is a good investment; it can contribute to both the economy and the efficiency of the building through its life. Good design can be realised by not only focusing on aesthetics but by paying attention to:

- Internal and external layout, with particular attention to efficient and economical use of space.
- Efficient use of energy, for example in the heating system (e.g. ensuring good insulation) and for the mechanical services.
- Low building maintenance costs
- Building and space flexibility, to allow for changing requirements.

The formulation of an accurate design brief and the development of design in strict accordance with that brief are key processes for the client or the project manager to oversee. In practical terms this means:

- Close collaboration between the client, the users and the design team during the development of the design.
- A procedure for formally checking that the developed design meets, but does not exceed, the requirements of the design brief.
- A formal presentation of the developed designs to the client and users.
- A formal 'sign-off' by the client that the design is correct and acceptable.

The client must indicate clearly the programme for project delivery and the design team should respond to this in terms of their strategy for meeting critical points in this programme. The client should not expect to consider the design in isolation but consider, for example, its relationship with the local neighbourhood. In this respect the client should expect to consider alternatives before committing to the 'signed off' design.

Through the project there will be incremental changes to the design caused by, for example, a change in the needs of the client, changes in technology or legislation. Formal processes for absorbing, reacting to and communicating these changes should be laid down and agreed by all parties.

On projects of sufficient complexity to necessitate the appointment of a project manager, it is unlikely that one person will have all the design skills. Special attention to communication and information flows is necessary where the collaboration of several designers is necessary, for example in specialist buildings or buildings that have sophisticated mechanical and/or electrical installations.

Stakeholders, other than those involved in building, are unlikely to be conversant with reading and interpretation of design drawings. It is important to ensure that the design team present their proposals in a form that can be readily understood. Pictorial views are useful here and modern technology (in particular computer-aided design) easily enables three-dimensional presentations.

Cost control overview

It is essential that the client understands the difference between:

- Estimating – the provision of an informed opinion at a particular time of what the final cost of the project is likely to be.
- Cost control – the management of the consequences of the design and construction processes so as to achieve value for money and ensure that the final cost does not exceed the budget.

Estimates are unlikely to be fulfilled unless cost control is exercised. Estimates provide a measure against which to control costs. They can also be used to assess project viability, obtain funding, manage cash flow, allocate resources, estimate duration and prepare tender prices. The techniques used to produce estimates vary according to the type and level of data available at the time of preparation. The accuracy of the estimate will improve as the design of the project develops and more details become available. The range of accuracy is likely to be in the region of $\pm$ 30% at project proposal stage to $\pm$ 2% at final tender stage. Providing cost control is being exercised, the accuracy of early estimates can normally be of sufficient precision for them to be valid parameters for decision-making and for the management of the project. A fundamental principle to be understood by the client is that projects can be designed to cost or controlled on the basis of costing the design. In the former a budget is set and the design is controlled such that at the final tender the design will reflect a cost which is, say, within $\pm$ 2% of the budget. In the latter the client understands that cost is not the primary driver.

For estimates to be effective, they should be supported by:

- A list of assumptions on which the estimate is based.
- A risk analysis – an identification of the potential risks together with an assessment of their probability, their likely impact (cost consequence) and the time at which they may occur.
- A sensitivity analysis – a statement of the comparative effects on the total estimate due to changes in principal data.

An estimate should include an amount to cover the uncertain cost of risks. This is known as the contingency, which should be sufficient to cover risks and is not eroded by any lack of cost control.

A particular concern of the client's may be the running costs of the finished project – a stance that is increasingly common. Higher initial building costs can result in lower running or maintenance costs during the project's life. The term 'cost-in-use' is used here to describe these estimated running costs and their inclusion in the estimates enabling the client to consider total project costs rather that just design and building.

Cost control procedures will be more effective the earlier they are instituted. By way of simplistic illustration:

- Cost varies with (but not in direct proportion to) size; once the size of a building is fixed, so is the general level of cost.
- The selection of the most economical design for basic elements such as foundations, structural frame, external cladding and roofing, is of far greater cost significance than the types of finishing.
- The overall cost of mechanical and electrical systems is largely governed by early decisions as to the environment to be achieved and therefore the type of system selected.

Methods used for cost control differ radically between the pre-construction and construction stages of a project. Cost control during pre-construction depends partly on the procurement strategy but more on control of the design process within that strategy. Pre-construction cost control, in simple terms, comprises:

- Preparing a cost plan – an allocation of the project budget into cost centres which match readily identifiable elements of expenditure, with allowances for, say, contingencies, reserves for changes in market prices, etc.
- Checking the cost of each element as it is designed against its

allowance in the cost plan. It is important to note that if this cost, as designed, exceeds its allowance in the cost plan, the excess can only be corrected by:
○ Redesigning the element to reduce its cost or
○ Transferring money into that element from contingencies or from another element yet to be designed
If neither is done, the control budget will be exceeded.

Successful cost control during the construction stage depends on avoidance of major change after commitment to build. Effective management here depends on:

- Design work being completed and fully coordinated before a commitment to construct is made (this applies in a progressive way when design and construction overlap).
- The contract being administered efficiently and promptly in accordance with its terms and conditions.
- Changes to the design being minimised after construction has started.

Regular cost reports should be produced throughout the construction stage of the project. From these, potential overspending can be identified before it occurs and corrective action taken. The client should, however, recognise that such corrective action is not always beneficial, since:

- It is likely that cost savings can be made only by reductions in standards or by omissions of part of the work remaining to be finalised – that is, the visible finishing and fittings. This is likely to result in the requirements of the brief not being fully met.
- Late cost savings are inefficient as any amount saved will be reduced or may be negated by the costs of disruption inherent in making the changes needed to generate these savings.

All estimates and cost control procedures should take inflation into account (i.e. the increase in construction cost between the date of the estimate and the date when the work is carried out). It is essential for effective cost control that:

- Allowances be made for inflation in all estimates and that the assumptions on which such allowances were calculated be stated.
- Such allowances for inflation be clearly identified within the estimates and not be used to correct other overspending.
- The assumptions on which inflation allowances are calculated be

reviewed regularly and as each new estimate is prepared the allowances adjusted in accordance with such reviews.

- A clear policy and method for drawing down inflation reserves be established and applied.

The client should distinguish between inflation and the effect on construction prices of market conditions at the time of tender. Contractors tend to adjust their tender prices by reducing or increasing their target profit margins in accordance with commercial factors. It is unwise therefore to base early estimates on any assumptions as to market conditions at the time of tender as the construction industry is subject to wide and comparatively swift changes in workload in accordance with economic conditions.

It can be expected that the project will not proceed as originally perceived or planned – the unexpected will happen. Such unforeseen happenings should be covered in the cost control system by contingency sums and the client should have a policy for the management of such contingency funds to ensure that, at all times in the project, the remaining contingencies match the remaining risks. The client may also establish and control a client reserve, not disclosed to the consultants or the contractors, to cover, for example, late changes in user requirements or unforeseeable third-party events.

12.6   Quality control overview

The quality (or performance) of the project, the completed building, is governed progressively by:

- The project brief – a clear statement of the standards of quality required. Great care is needed to ensure this standard is unambiguous; such phrases as 'the highest quality attainable within the control budget' are to be avoided.
- The original design, which must include:
  - The adequacy of the components selected for the building
  - The interface between related components and systems
  - The integration of mechanical and electrical systems into the overall design
  - The completeness of design before construction starts
- The specification in terms of both:
  - The conversion of the quality standard demanded by the project brief into precise requirements for both the supply and the installation of materials, components and systems.
  - The setting out of criteria against which the standard of the

finished work will be judged, for example by reference to standards, codes of practice or similar.

- The quality control system – control mechanisms which can be applied to the execution of the work on site; for example the detailed ongoing supervision by the contractor, the programmes for testing, the procedures for rectifying any defective work.
- The inspection systems – the independent inspection and verification of the contractor's work by the design team. This includes procedures for witnessing tests, for commissioning plant and systems, for pre-completion inspection for defects lists and defect rectification and for final hand-over.

The client may choose to appoint a quality inspector (or a clerk of works) whose sole function is to inspect completed work and verify it against the imposed quality standards. However, such an inspection and verification regime should be recognised as a last line of defence. The key to quality control on site is to specify clearly and to monitor closely the quality control activities carried out by the contractor while work is being done.

Many construction companies and firms of consultants within the building industry have adopted, or are in the process of adopting, formal quality assurance systems for their own work or services in accordance with BS 5750, ISO 9000 series or equivalent standard. These organisations are better placed to understand and to delivery required quality products and performance.

12.7   Change control overview

The management of change should be a prime objective of the project strategy. Clients should not expect to make changes to 'signed-off' and agreed designs as such changes can be the cause of significant extra costs and delays in construction projects. However, circumstances (such as the economic environment, the client's needs) do change and so change to a project may be unavoidable, therefore alterations should be understood and dealt with in a previously specified manner. Consistency of approach is vital; it helps to ensure transparency, effective communications and eventual project success.

Client changes (as distinct from design development) are changes made by:

- The client after the design brief has been agreed. The most significant of these are changes to the scope of the works.
- The design team, with the approval of the client, to a design

feature which it has been decided should be frozen, or 'improved', or altered to overcome design errors or inconsistencies.

Client changes can be relatively minor, such as adding a few extra power points, or they can be major, with major cost implications, such as the addition of an extra storey. The cost of client changes depend largely on when they are made; for example:

- Before the construction contract has been let, the cost can be contained to that of the changed feature itself, and perhaps some nugatory resource cost.
- After the construction contract has been let, the cost will probably be disproportionate to the value of the change. It can disrupt the contractor's work and invariably gives rise to a higher cost than if the change had been included in the contract as let.
- Many minor changes can have a cumulative and serious effect: disruption claims caused by a large number of small changes are common.

Some client changes are unavoidable. Examples of such changes are:

- Compliance with changes in legislation.
- Requirements of the health and safety or fire prevention authorities.
- Those required by unforeseen ground conditions.
- Previously unforeseen users' requirements.

The contingency in the project budget should be sufficient to take account of the likelihood of such changes based on risk assessments.

Changes proposed prior to construction may be either unavoidable or optional. If they are unavoidable, the client should authorise a transfer from the contingency in the budget to cover them. If they are optional, the client is advised to approve them if it can be demonstrated that they offer good value for money (or a saving) and that there are sufficient funds available to pay for them.

Changes proposed after the construction contract has been let can have major time and cost implications and should be avoided if at all possible. If they are not essential, it is recommended that they should be deferred until the project is complete and then reviewed to see if they are necessary and economically justified. The need for changes should be minimised by:

- Ensuring, with the assistance of an experienced project team and designers, that the project brief is as comprehensive as possible and the users have 'signed it off'.
- Taking into account any proposed new legislation.
- Having early discussions with outside authorities so as to identify their requirements.
- Undertaking adequate site investigation or condition surveys if existing buildings are to be renovated.
- Ensuring that designs are fully developed and coordinated before construction contracts are committed.
- Imposing discipline on users to finalise and 'sign-off' their requirements in strict accordance with the project programme.

When change is suggested, the client should always call for:

- The reasons for the change.
- The source of the change.
- The full cost and time consequence of the change.
- Proposals for avoiding or mitigating time overrun.
- Source of funding of any cost overrun, e.g. contingencies, client cost reserves, and/or compensating savings elsewhere.

12.8   Commissioning

Once the building work is complete, the systems which will support user comfort (e.g. heating, ventilation, and lifts) must be commissioned to ensure they are working effectively and reliably. In relatively simple buildings, the client can insist that this is a function the contractor must perform. Where buildings have sophisticated systems controlling the internal environment or facilitating staff movement or safety, commissioning can be established as an independent activity carried out by specialists. However commissioning is facilitated, it must be complete before any building can function effectively and be handed over.

12.9   Occupation and take-over

The client is responsible for addressing the issues of occupation, staffing and subsequent operation and maintenance of the building. This activity is separate from the design and construction process, although it will affect it, and will have its own time, resource and cost implications that should be incorporated into the overall project plan. For large projects, arrangements for the nomination of a member of part of the client's organisation to act as occupation manager may be made to manage this activity or the client may appoint an external facilities manager.

Occupation plans should be established during the design stages of the project and should cover:

- The operation of the building on a regular ongoing basis
- The hand-over and acceptance of the building from the contractor(s)
- The progressive final fitting-out (if any)
- The physical occupation of the building with minimum disruption to the client's operations.

12.10   Conclusion

This chapter discusses the management of the project and has frequently referred to the project manager. Whilst there is a growing appreciation of the need for a member of the project team to manage the process and the appointment of a project manager is increasingly common, it is possible for the management role to be achieved without this appointment. However, the fragmentation of the design and production processes makes the task of project management very difficult for a member of the design or construction team to carry out as well as their primary role. Managing the realisation of the product is a key role in achieving the client's aims and a separate appointment is advantageous, particularly to less experienced construction clients.

References

Chartered Institute of Building (1996) *The Code of Practice for Project Management*. Englemere Services Ltd.

Construction Industry Board (1997a) *Briefing the Team*. Thomas Telford Publishing.

Construction Industry Board (1997b) *Code of Practice for the Selection of Subcontractors*. Thomas Telford Publishing.

Construction Industry Board (1997c) *Code of Practice for the Selection of Main Contractors*. Thomas Telford Publishing.

Construction Industry Board (1997d) *Constructing Successes*. Thomas Telford Publishing.

Construction Industry Board (1997e) *Liability Law and Latent Defects Insurance*. Thomas Telford Publishing.

Construction Industry Board (1997f) *Partnering in the Team*. Thomas Telford Publishing.

Construction Industry Board (1997g) *Selecting Consultants for the Team: Balancing Quality and Price*. Thomas Telford Publishing.

13 Facilities Management

John Hinks

13.1 Introduction

This chapter critiques the limitations of current approaches to measuring facilities management (FM) performance, and notes the tendency to emphasise FM inputs and outputs. It discusses the complexity of measuring the value of the contribution of FM to business – an issue more related to business outcomes than FM outputs.

The chapter suggests that more flexible and bespoke sets of performance measures are needed to allow performance assessment to illuminate the true contribution that the FM function makes or could make to business. It also suggests that these assessments need to be multi-faceted and relate to the context of the specific organisational circumstances. Performance measures should reflect the context and priorities of the specific functional need of the organisation which the FM function is supporting. They should therefore be developed in conjunction with the business customer, so that they adequately correspond to the range of outcomes and performance drivers that underpin the effectiveness of both the organisation and its facilities management.

A review of the literature and published practices on FM performance indicates that much of the current emphasis of FM performance measurement appears to be driven by the general comparability of the indicators *per se*, most of which are facilities-oriented and not business-outcomes-oriented. There also appears to be a general preference for assessing performance using quantitative measures. This concentrates attention on the lowest common denominators of FM, not on its distinguishing features, which also pre-loads the analysis of performance onto the operational side of FM at the expense of the strategic. It also inhibits the realistic assessment of the usefulness of FM to the particular business.

Because of this, the most appropriate indicators of FM performance will vary considerably in content and/or priority according to the particular business circumstances. Measuring the usefulness of FM for particular business needs will need to be driven by a bespoke and organisationally-oriented approach. There will remain problems of identifying, categorising, and assessing the business outcomes of FM in their local context. The chapter presents examples to illustrate this. Notably, the terms of FM performance assessment must be expanded from the current approaches and indicators to also include those which correspond to outcomes and performance drivers of effective business. These will be very particular to the circumstances.

13.2   Contemporary approaches to assessing FM performance

To date, most of the publicised attempts at measuring FM performance have been benchmark-oriented and have focused on using *quantitative* assessment (McDougall, 1999). Usually these approaches involve making comparisons between distinguishable and measurable dimensions of the facility or the facilities function. Hence, and in the absence of other, more easily-assimilated, measures for the objective internal assessment of FM performance, facilities managers have looked comparatively towards their compatriots, particularly those who appear to perform similar tasks, use similar procurement arrangements, operate in similar business environments, and/or simply have similar building stock. The assessments are usually made from within the FM field, using FM terminology and FM priorities, and are primarily made for consumption by facilities managers.

The comparisons appear to be usually used by facilities managers to check their efficiency with that of their competitors. Obvious difficulties arise with this approach where FM differs locally in its definition, role, procedures and/or environmental context. In addition to this and other hidden differences and/or unfathomable idiosyncrasies in data, there may be other reasons to doubt its direct comparability, including the openness and reliability of the stated figures. One of the lessons that has consistently emerged from comparing superficially comparable quantitative data is that the comparison of performance measures has to respect their origination *and* their application. This does not necessarily correspond with the actual nature of the contribution FM makes to individual businesses, or to the multi-faceted nature of its involvement with the organisation (a feature of FM that appears to be poorly understood also).

However, looking at the normal practices in FM performance assessment, the tendency to isolate the FM functionality from its context in order to use comparative single-dimension measures reduces the performance data to measurements of the lowest common denominators. This approach in itself circumvents much of the potential value in the exercise, and also does not readily support the appraisal of the integrated contribution to business. In other words, the performance assessment agenda becomes driven by the availability of quantitative data, not by the original need to assess the real contribution of FM to business.

So, in general, attention has not focused enough on measuring the value of the integrated FM function as a whole, or on the quality of the alignment and interrelationship between this integrated function and the particular business support needs of individual businesses. To paraphrase Robert McNamara's hard-learnt observation after trying to measure the success of the war in Vietnam, 'the attention has erroneously fallen on the importance of measurement, not on the measurement of the important'. To date, the consequence of this for FM appears to have been that the design of performance measurement tools, and the resultant search for generalisable key performance indicators and data, have tended to focus on the quantitative measurement of base FM inputs, elements of the FM process, or, at best, FM outputs. As McDougall observed (1999), 'the evolution of performance measurement of Facilities Management has progressed towards generalisation'.

Many of these measures are activity-oriented or concerned with how FM efficiently performs single-issue tasks, not with the real importance of FM, such as its usefulness to the needs of a particular business.

Bearing in mind that integration was supposed to be one of the main (business) drivers for the emergence of facilities management, this represents a potentially serious mismatch between the nature of the FM role and its assessment. If this preoccupation with the parts rather than the sum of the parts continues, such an approach to performance assessment could do FM generally an immense disservice as the business focus evolves explicitly towards the complex issue of 'best value'. Unattended, the continuation of these 'piece meal approaches to the performance assessment of FM' (McDougall, 1999) may do more harm than good – to the business as well as FM.

13.3   Issues in contemporary FM performance assessment

In practice, it appears that many organisations or industries develop their own idiosyncratic key performance indicators. Since FM spans so many industries it makes it very difficult to envisage a generalised set of performance indicators, which supports performance assessment beyond the basic, widespread, and quantitatively measurable activities. Herein lies a dual risk: first that the comparability is rendered superficial (by its generality); and second, that the extent and relevance of the degree of this superficiality is itself obscured too. It also makes it very difficult to introduce more flexible and encompassing developments in performance assessment.

Choice of indicators

However, adopting this generalising approach raises another compounding problem for performance measurement, and one that is more fundamentally significant for FM. Most of these types of indicator provide data that are applicable with interpretation and intuition by the expert to monitoring the management of the facilities management tasks. However, the business customers for FM may not see the indicators as relevant. As Dixon *et al.* (1990) noted in the context of managing performance, the value of performance measurement lies in enabling organisations to aim their actions at achieving their strategic objectives. Linking this to another fundamental problem – that FM does not generally contribute to organisational strategy formation – this may mean that many of the FM performance measurement tools in current use are really only suitable for an internal audience involved in the optimisation of an operational service function. In retaining the sole focus on this, performance assessment risks polarising attention away from that which business managers consider to be strategically relevant.

Meanwhile, for many FM organisations the issue remains one of survival in a turbulent procurement market driven by the tension between in-house and outsourced provision of the FM service. Until recently, cost reduction seemed to be the overriding criterion for making procurement decisions, and so it is understandable that there has been an historical tendency to focus on financial measures as the principal performance indicators of FM. But as a more introspective view on the drivers of organisational competitiveness emerges in business, the primacy of cost minimisation is being challenged by the best value initiative for example. Facilities managers need to take the opportunity of the broadening agenda to relate FM to assessments of business outcomes and performance drivers accordingly. Consider the BAA viewpoint, a client that has rapidly become an icon for client attitudes towards the construction and other service industries as a whole. 'Identify what is important to your customers; formulate these issues into key performance indicators; set these as company targets; and then, measure achievements over time.' (Van de Vliet, 1997). Is contemporary FM performance assessment really addressing these issues?

The business perspective

Whilst Massheder and Finch (1998) suggested that specific metrics are essential for ensuring a common understanding of performance and to identify performance gaps, they concluded that FM metrics were not being used to reflect the overall performance of the business. Their conclusion was stark: 'The significance of facilities decisions on overall business success will continue to go unnoticed until a more holistic approach is applied to benchmarking.' The significance of this reinforces the comments made by Van de Vliet (1997), who noted that, in the context of customer satisfaction, it is especially important to recognise the importance of particular issues to your customer, and to assess the service against these needs. Whitaker collected the issues together neatly, and as long ago as 1995 observed that:

> 'more than ever, senior management is scrutinising the contribution of facilities management departments to an organisation's overall success ... to improve, facility managers must first understand how their group is doing – are they meeting the expectations of management and their customers? ... facility managers who do not know the answer to that question will have a difficult time documenting their contributions and improving their services.'
>
> Whitaker (1995)

If the business view of FM is to develop to recognise the strategic potential of FM, it may have to be led by a corresponding broadening of what facilities managers view as constituting performance.

Performance metrics

Hence the issue arises of what forms of metric should be used. Vorkurka and Fliedner's paper (1995) was one of several to identify that, in the pursuit of performance measurement, a balance between financial and non-financial measures is desirable. This served to reinforce Becker's (1990) approach to performance assessment that encompassed a collection of indicators that would be relevant to the business-oriented customer. By 1998, Vorkurka and Fliedner (1998) were discussing the critical distinctions between agility and flexibility, and noting in connection with this the need 'to build cumulative and lasting improvements in competitive capabilities and provide a base for developing agile practices and competence'. Whilst their message was aimed primarily at the general business manager in the wider organisation, as an integrative component of that larger organisational system, facilities managers should take heed, too. O'Mara *et al.* were quite categorical about the need for this:

> 'Performance measurement systems not only provide the data necessary for managers to control business activity, they also influence the behaviour and decisions of managers. This being the case, a restrictive set of financial performance measures may adversely impact on that organisation's long-term viability, so organisations should develop a broad range of performance measures.'
>
> O'Mara *et al.* (1998)

Could this have already happened for FM?

13.4   Assessing the integrative nature of FM performance

Consider also the need to accommodate the integrative nature of performance drivers (Leaman,1995). Going back to Becker again (1990), his approach broke new FM ground in presenting his range of performance assessments graphically, and as a comparable performance profile. This was designed to allow comparisons to be made between desired and actual FM performance, an approach that made it easy to communicate how well the facilities management department was doing and where it needed to improve. Through this, Becker (1990) explored the types of performance profile that would correspond to the range of permutations of service levels and cost which different customers with varying business support priorities would need. He produced a set of profiles, differentiated according to the priorities of the customer, and which co-related similar issues to those discussed by Vorkurka and Fliedner (1998) – for example, cost efficiency, dependability, quality, and flexibility of the FM service. Varcoe (1996) echoed this broader combination of issues in the conclusion to a paper on business-driven facilities benchmarking: 'quality and overall performance must be on the agenda as well as cost'. Surely in this context the overall FM performance becomes absolutely specific to the context of its business application and cannot be sufficiently assessed other than by the inclusion of some reference to the weighted priorities of the specific organisation under focus. If so, the assessment criteria need to be designed as a bespoke tool, and set within a framework which is compatible with the business view of its performance drivers.

Becker's (1990) approach probably remains one of the best

philosophically. He illuminated the need to consider the design and use of performance measurement tools in the context of the specific organisation. He did not explicitly model the integrative issues of FM in his tool however (Leaman, 1995), and this is probably its major limitation. But like Vorkurka and Fliedner (1995) later, he understood that no single measure can adequately provide a clear performance target.

Data issues

Returning to Becker (1990), another strength was that his model operated on a relatively small number of performance indicators, reaffirming the suggestions of Slater *et al.* (1997) that performance assessment frameworks should be limited to between seven and 12 performance measures in order to be manageable. There still remained the problem of data availability for these measures, and, like the other performance assessment models for FM, Becker's (1990) approach concentrated on the operational FM issues for which the FM had plentiful data, much of which can be quantitatively expressed. Most of the data were also historic and summaries however, so like other assessment tools, in its 1990 form the model remained a 'snapshot' approach. This fundamentally limits the scope for assessing changes in performance over time, which as a continuous improvement feature of an FM service level agreement could be the major business customer concern in performance assessment and moreover could be an issue that they qualitatively assess.

Further evidence of the value in Becker's recognition of the need to look beyond the quantitative is borne out by analyses of performance assessment from other disciplines. Consider Chakravarthy's conclusions in 'Measuring strategic performance' where he sought to identify the appropriate performance measures for distinguishing excellent firms in the computing sector:

'no single profitability measure seems capable of discriminating excellence. Moreover, accounting data that are typically used to construct these measures capture past performance or historical trends. Strategic performance needs a more futuristic measure.'

Chakravarthy (1986)

The historic and periodic value of summarising performance for the purposes of external financial appraisal against traditional accounting criteria was also clearly recognised by O'Mara *et al.* (1998), but so too was the need for this to be used as part of a wider set of performance measures. This was not a new call: others had identified the need for a broad range of performance measures, including Fitzgerald *et al.* (1991), and Kaplan and Norton (1992) in their discussions on the very popular balanced scorecard. Furthermore, Keegan *et al.* (1989) predated both of these by looking at the issue of obsolescence in performance measures.

Adjusting the assessment

Chakravarthy (1986) also raised another performance assessment challenge to consider during the transition between 'cost-minimisation' and 'best value,' that of recognising the need to retain the scope to adjust performance when assessing performance. This is an issue which, at the time of writing, does not appear to have been considered in the FM performance assessment literature. As Chakravarthy concluded:

'a firm is excellent only if it has, in addition [to satisfying direct financial criteria], the ability to transform itself in response to changes in the environment ... [These] criteria ... are necessary conditions for excellence, [but] they are not enough ... Excellent firms in our sample were able to generate more slack resources than non-excellent firms... not all the slack invested by a firm can be quantified.'

Chakravarthy (1986)

Another issue emerging here is whether organisations are applying performance measures to help distinguish the capability of their facilities management service to evolve, and/or the value of the FM during the process of business change. Also implicit within this is the need for the currently used performance indicators to be easily and incrementally adjusted without affecting the use of the tool.

In summary, it remains difficult to make faithful and representative assessments of the business outcomes of facilities management, principally because the assessment agenda is not wide or deep enough. This matters because whilst consolation continues to be sought in the comfort of detail, the emphasis on generalised solutions to performance assessment directs attention towards the less competitively distinguishable facets of modern facilities management. This indirectly helps prolong the overlooking of the strategic potential of FM by business. Furthermore, as the industry reinforces its commitment toward creating generalised FM data sets through various cross-industry and international benchmarking initiatives, so the demonstration of the individual and specific value of FM will have to be resolved by alternative and bespoke performance measures, and will have to be done locally by individual facilities managers. This raises separate issues of rigour and internal consistency in assessment. However, there is much at stake for the individual FM service. This is an issue that cannot be ignored, and the remainder of the chapter explores the challenge that this places before the FM profession.

13.5   Defining FM and the nature of its contribution

The lack of a clear and defined recognition of FM by the business customer base is part of the chronic nature of the problem of assessing FM performance. This is another dimension of the problem with measuring performance arising from the essentially variable relationship between FM and business. It is also further evident that some (or most) of the content of a performance assessment tool needs to be bespoke. Ultimately, performance assessment remains a vehicle to better understanding of the fitness of the service, but in the meantime as the assessment techniques develop they have to be sufficiently flexible to help the business customer and facilities manager understand and appraise the usefulness of FM to the individual business. This is something that initiatives to date have indicated should be done by facilities managers and business managers together. As O'Mara *et al.* (1998) concluded, there remains an unsatisfied need for a formal recognition of the link between strategic management (including strategic change) and performance measurement. They found no formal linking of the two, and at best key performance outcomes, rather than key performance drivers, would determine the effectiveness of actions connected to strategic decisions. Tranfield and Akhlaghi encapsulated this in their paper on relating facilities to business indicators:

> 'In the context of the whole organisation, the role of facilities management has gradually evolved from merely helping the organisation to survive to acting to enhance its potential to prosper in a volatile commercial climate. It then follows that the challenge for facilities managers is indeed the same challenge facing the organisation.'
>
> Tranfield and Akhlaghi (1995)

Business outcomes

Given this, the criteria for FM performance assessment should be compatible and coordinated with the organisation. The achievement of this requires attention to be refocused on the high-level outcomes from FM which matter to business managers, measured dynamically to allow past changes to be evaluated and future changes to be translated from business level strategic priorities into FM strategies. Furthermore, this has to be correlated with the drivers, which can be used to create performance improvement. Bear in mind that, in the broader changing business environment, improvement includes adjustment of the existing service to maintain a good fit with the changing (and changed) needs of the business. Sometimes this involves reaction, other times proaction and leadership. Hence performance does not necessarily come down to efficiency alone. As Tranfield and Akhlagi have pointed out:

> 'Facilities Management and facilities services are often locked into the cost-cutting and efficiency-drive paradigm which makes any attempt to compare output and especially cost parameters to be seen as a negative tool to reduce budgets and expenditures even further.'
>
> Tranfield and Akhlagi (1995)

Weighting and emphasis of indicators

There is an additional consideration to this broader interpretation of the use of measures from the outset. Where the strategic priorities of the organisation require a qualitative and integrative service provision, the measurement tool and the metrics must be sufficiently sophisticated to allow the revised targets to be expressed in terms that can be weighted and assessed flexibly. For this to occur, engaging the business customer in the design of business-oriented FM performance assessment criteria becomes axiomatic. As O'Mara *et al.* (1998) observe, there is also a need to engage the industry-level service practitioner, too.

Most of the generally recognised FM key performance indicators emphasise facilities-oriented issues. This may be because they are designed usually by facilities managers for facilities managers. This approach tends to involve using imported and generalised sets of indicators for benchmarking the common activities of FM between organisations. But when these indicators are scrutinised for their usefulness to individual businesses they may actually prove to be incomparable. Even at the level of the particular FM application, as McDougall (1999) observed '[FM] definitions often mean different things to different parties, between organisations, or even between departments.' Making the step to using bespoke sets of performance indicators and measures that correspond to the localised business priorities for FM will require a sea change in approach.

13.6   Issues for benchmarking FM performance

Benchmarking which uses a restrictive set of indicators based on existing practices will still only allow an assessment of the efficiency of comparable processes. Over time, focusing on this may have a tendency to normalise rather than differentiate the focus on FM. Clearly, some intuition and good sense is required to interpret, extrapolate, and act strategically upon these results at the level of the individual organisation. This is where the limit of current benchmarking emerges, especially taking into account the variability in local definitions of FM. Another issue is that benchmarking over time, which is essential to establish continuous improvement, tends to require a freezing of the criteria for performance assessment, whilst the functional requirements or priorities may move on. The contribution of benchmarking lies more in eliminating relative noncompetitiveness in FM rather than offering relative FM competitiveness to business. As will be discussed later, the two are not quite reciprocal. By association, perhaps benchmarking is also more applicable to processes undergoing little or slow change rather than processes undergoing rapid competitive evolution. However, achieving competitive business advantage probably emerges from adaptability and differentiable forms of performance, perhaps by doing something differently rather than simply more efficiently. These facets of business process effectiveness may be obscure to superficial benchmarking, an issue of 'apples and oranges' as Scott (1998) put it.

Summarising the issue of performance measures and benchmarking, the effect of the growth in a reductionism approach to FM assessment has been to turn assessment away from the bespoke nature of the FM role in business support. Additionally, rather than simply demonstrating its contribution to avoiding noncompetitiveness, FM has to demonstrate the value of its contribution to business competitiveness. Both are needed, preferably as a combined assessment. The focus of FM performance assessment must be expanded to the measurement of its usefulness to the individual business. However, individual assessments need individual tools and measurements, which raises the problem of identifying and representing the local business perspective on FM outcomes and contribution to overall performance.

13.7   The complexity of measuring the usefulness of FM

In order to be realistic and relevant and be perceived as such by all involved, performance indicators should be derived by consensus between the FM service provider and their business customer. Accordingly, if the usefulness is to be assessed, then measurement of FM performance should be based on a combined assessment of facilities and management outcomes.

However, at the level of its role in the overall business process, the value of FM is immensely difficult to disentangle from other contributory factors. The problem has three facets, at least. First is the tendency to focus measurement where the distinguishable data are available and, ideally, are quantifiable too – as discussed earlier. The second facet is the consequent reduction of the FM service into discrete key performance indicators that say little about the collective contribution or pertinence of the FM service for the individual business. Indeed there are indications that business managers simply do not consider many of the key performance indicators in common FM use as relevant. The third facet is the tendency for FM performance measures to be facilities-oriented and inward-looking rather than corresponding to what the business may or may not recognise it needs. In contrast, a review of business texts indicates a concern about value drivers which bears little or no correlation with the measures which FM performance assessment focuses on. These measures are also frequently qualitative in nature, tend to change over time, and are expressed in business language.

One of the key issues in making a realistic performance assessment is the integrative nature of FM. Usefulness in FM is conferred by the alignment between the highly bespoke and ideosyncratic combinations of individual facilities management services and the core business process. The effectiveness of this alignment is itself subject to the changing balance and interaction within the FM role and between FM and the organisation. In this context, FM can be both driver and recipient of changes in the interrelationship – usefulness and effectiveness are measures of outcomes of that dynamic relationship as well as the inputs. Accordingly, they change over time, too, as both the priorities and the nature of business support needs change. So, efficiency of these operations and their effectiveness, individually and in combination, all need to be measured from a joint perspective.

Of course, effectiveness is much more difficult to define and disentangle for measurement purposes than the efficiency of inputs from isolated activities – not least because value and effectiveness depend on the synthesis of many contributions and usually appear relatively remotely from where and when the individual inputs occurred in a dispersed form throughout the value chain. The difficulty of disentangling the individual contribution of a particular service can confound the definition of assessment criteria and metrics. Also the very process of doing this tends to strip away the essence of the integrative role of FM in the business process. It also seems probable that disputes over the criticality of a particular role will detract from the impartial measurement of its partial contribution to business outcomes. Furthermore, as referred to earlier, different service-based or process-based businesses will need different types of FM service. This will implicitly weight the various components of those services differently. Furthermore, these weightings will change over time, and may even differ concurrently in the context of competing strategic business priorities or even different sub-sectors of the business. This situation is going to have to be addressed if FM is to be recognised for its potential and actual value to business.

13.8   Moving from measuring FM outputs to evaluating business outcomes

If the performance assessment of facilities management is going to shift successfully in emphasis towards the usefulness of FM to the business, there needs to be a more overt recognition of the relevance of the business environment to the role of FM. Consider the following hypothetical cases of the same business facing different circumstances. These examples illustrate how different dimensions of the FM role have to be assessed differently in serving the ongoing, relatively stable needs of business or alternatively how well FM prepares the business for changes in its priorities or its operating environment.

Scenario 1: FM in an environment of business stability

For an organisation operating in a stable business environment, the success criteria for FM may relate to reliability of service quality and a dependable speed of response set against continuous improvements in cost efficiency. Note that these measures still represent *FM outputs* however, and not business outcomes – see scenario 1 on Fig. 13.1, noting that the poles refer to FM outputs and not to business outcomes. But in taking this approach to measurement there is a risk for FM. For whilst there is a superficial attractiveness of the relative ease of benchmarking a stable function, the focus can settle on making comparisons between process efficiency and input/output ratios. In this circumstance FM may become attuned to serving a stagnated business scenario, and lose its ability to support changes if or when they are needed. Hence a second dimension of FM needs to be considered.

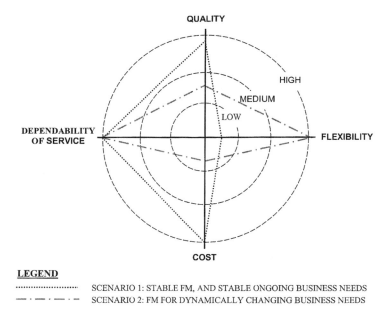

LEGEND

.... SCENARIO 1: STABLE FM, AND STABLE ONGOING BUSINESS NEEDS

— · — · — · — SCENARIO 2: FM FOR DYNAMICALLY CHANGING BUSINESS NEEDS

Fig. 13.1   A four-pole diagram for 'FM outputs' driven facilities management in various scenarios.
Source: Developed from Fig. 7.10, in Scott, M. (1998) *Value Drivers*. Wiley, p. 105.

Scenario 2 – The role of FM in achieving business flexibility for competitiveness in a dynamic market

This is a more realistic scenario for most businesses. The value of FM arises from a combination of its support for the ongoing needs of the business and its support for changing business needs. Key observers of business competitiveness are remarking that speed and appropriateness of change are becoming increasingly relevant elements of competitiveness in a global business market. In such circumstances, it is reasonable to assume that the future competitiveness of a significant proportion of the business world will become increasingly dependent upon dynamic competitiveness – that is, agility and competitive change in business processes (and products). In such circumstances, relative business advantage in a dynamically competitive sector is likely to be derived from the strategic application of FM in a customised manner. The indicators for assessing this aspect of FM performance will be have to be more high-level, more transparent to business, and will tend to represent the integrated output of a range of interrelated FM services, probably the result of a complex interactivity within and with the core business. They are likely to be particular to the business needs of today.

So, these success criteria for FM are more likely to be associated with innovation. High cost may matter less than speed of delivery measured in terms of flexibility. Here the resource implications are more open-ended as the service needs are less defined and stable. Cost-based or time-based metrics will not be sufficiently illuminating if used individually, and it may be that a combination of metrics needs to be looked at as a suite in order to appraise the multidimensional nature of the FM contribution. This represents a bespoke attunement of the FM service to the user's individual competitive needs. In industries where distinctiveness and agility of process is an element of competitiveness, this could be where the distinguishable value of FM lies – see scenario 2 in Fig. 13.1, noting the relatively low emphasis placed on cost efficiency.

Repeatability and consistency may be less important dimensions within the assessment of FM quality than the degree of fit of the service to a varying business need. This places the emphasis on FM fundamentally in the arena of business innovation, where the predominant features of good FM may major on adaptability, novelty, support for new processes, and/or timeliness. This situation differs considerably from the first scenario. Note how the FM output indicators used for scenario 1 now need some further qualification in their interpretation before the assessment of FM performance can be compared.

Scenarios 3 and 4: FM for survival in a dynamic market

To extend this further, imagine the market for FM within the dynamically changing business sector in scenario 2. The performance of this type of FM will probably be more intricate and therefore more difficult to assess, yet more relevant to the future of global business competitiveness and global FM. Consider then a third scenario where survival of the organisation depends on agility rather than cost-efficiency, as appears to have been the case recently in some of the electronics fields. In these circumstances, the target business value of FM may reside in its efficacy or usefulness when applied strategically for competitive business advantage based on change. Here the key performance indicators for FM may have to be more oriented towards business outcomes. The key performance indicators could converge with core business performance indicators such as agility rather than flexibility, business continuity rather than FM dependability, fitness for purpose as a qualifier of quality, or the ability of the service to contribute to changing the business (Scott, 1998; Vorkurka & Fliedner, 1998) – see scenario 3 in Fig. 13.2, noting that the poles in this figure relate to business outcomes, not FM outputs. In such circumstances, the FM output measure of cost (see Fig. 13.1) could be of minimal concern within the bigger

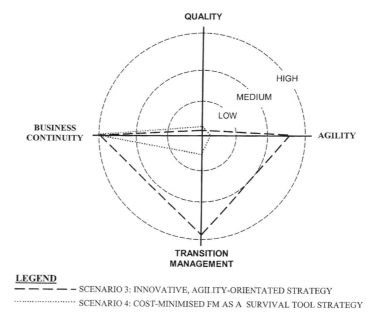

— — — — – SCENARIO 3: INNOVATIVE, AGILITY-ORIENTATED STRATEGY
·················· SCENARIO 4: COST-MINIMISED FM AS A SURVIVAL TOOL STRATEGY

Fig. 13.2 A four-pole spider diagram for business outcomes in various survival strategies.
Source: Developed from Fig. 7.10, in Scott (1998) *Valve Drivers*. Wiley, p. 105.

strategic picture of the overall business outcomes. In this scenario the emphasis moves from efficiency and output to effectiveness and outcome, and the possible needs are measured using a different set of indicators than the stable needs. Once again, attempts to make performance assessments using generalised and reductive FM indicators will not be especially valuable.

Finally, consider a contrasting alternate survival strategy in the same circumstances, which may hinge around the reduction of all costs to an absolute minimum at the opportunity cost of flexibility and dependability, and with no intention to manage transition, merely to streamline the existing operations – scenario 4 in Fig. 13.2. Here cost becomes the predominant measure of FM outputs and is likely to be the performance driver affecting the business-level outcomes. Achieving FM performance may involve shedding facilities which were highly efficient when measured against generalised criteria. In contrast to scenario 3, the emphasis moves from effectiveness and efficiency to cost minimisation. Note the equally high priority placed on the business outcome of business continuity in both scenarios, but the implicit need to consider the nature of the FM outputs differently in each scenario. So given any of the above scenarios, the FM relevance to the business will differ significantly, as should the assessment criteria for evaluating performance.

13.9   The challenge of attuning FM evaluation to changing business circumstances

The question of the variability in organisational capability to properly exploit the potential of FM has rarely been discussed. It seems reasonable to assume that variations in the nature and sophistication of business processes will give rise to differing needs from facilities management, and hence different understandings of performance and contribution. In these circumstances, any mismatch between the FM potential and the organisation process maturity could be a central factor in the appropriateness of FM performance assessments.

For example, a core business process that is essentially chaotic and poorly managed, that may be characterised by poor strategic planning and a high (and uncontrolled) variability in their business processes, will not be able to take advantage of stable, predictable, or change-enabling FM services (scenarios 1 and 2 earlier). In contrast, the sophisticated and highly process-driven business that is at a highly evolved level of business process capability will be able to take advantage of a highly sophisticated FM service. Each business will need an FM service that is highly attuned to its business capability and process. A mismatch in either circumstance could be damaging. The measurement criteria will differ in each circumstance, perhaps alternating between business outcomes and FM outputs perspectives as the business transcends from one capability level to another. For an FM provider servicing a range of organisations spanning these and other intermediary levels of business process capability herein lies another type of performance assessment challenge – to differentiate and optimise their service provisions to each organisation based on its capability to use them.

Contrast, for example, what would constitute usefulness or efficacy of the FM requirements of a business that was restabilising immediately after changing its core process/production. Or a business that was in the act of changing its core process/production (or indeed, considering the options), or was planning the change process, or has a stable core process which it simply wishes to continue supporting. Finally, consider the priorities for FM of a business that wishes to protect the consistency of its core process/production from external pressures by the creative adjustment of its support service.

Returning to scenario 1, which was the example of the business which is operating a stable process, and which in an unchanging state may be able to make most use of an operationally efficient and cost-minimised form of FM. Facility usage and demands on other aspects of the business support services may be stable and align well with the operational and facility-oriented face of the FM industry. The FM needs may be predictable and stable. Little investment may be planned for the purposes of change, and a reactive rather than proactive FM with little strategic input to the business may suffice and represent best value in these circumstances. For this type of organisational situation, the performance indicators may remain broadly the same as they are now.

Scenario 5: Strategically planning change

Next consider the same business in the options-analysis stage of planning a change in their core process or production/service provision. Here the adaptability of the core processes may be more critical to the future competitiveness of the business than the actual cost of the FM needed to ensure that provision. Proactive FM foresight and a strategic understanding of the interrelationship between the business and the FM may be very valuable to the business. Any stagnation or unresponsiveness caused directly by the nature of FM provision and outlook could severely limit its usefulness to the business during the change phase and the changed phase. Such unresponsiveness may limit the change options, or potential speed of change, and the efficiency of change *and* the changed business. It may dull the organisation's ability to respond to a turbulent environment. Note, however, that in addition to any problems with the inherent capability of the FM provider, such limitations may derive from the nature of the contractual arrangements for the provision of the service, and/or the remoteness of the FM service from the core business. Clearly then, the measurement criteria and key performance indicators would differ for the same business and same FM organisation in these two scenarios. They would also require to be highly contextualised for a business to be able to strategically optimise the FM usage, yet alone to be able to compare the *outcomes* of FM performance with others.

Consider now the issue of the adaptability of the FM function *per se* – this is something which applies to all of the scenarios discussed above (even scenario 1, since otherwise the FM function could be a limiting factor in making any business changes that might be needed in the future). The adaptability of the FM function may be crucially important for an organisation that is trying to protect the stability of its core process or production via changes in the nature of the facilities it uses for this. This is a situation facing many organisations in global markets as they downsize or rationalise their global facilities needs. The adaptability of the FM service may also be absolutely pivotal in the post-crisis adaptation of the core business process. Here cost efficiency is likely to defer to other factors such as operational effectiveness, the robustness of the transitionary or new service provision and alignment, and the quality of FM service alignment with the restabilising needs of the core business. Again, the indicators need to change accordingly.

Inevitably, the balance of importance, and hence the priorities for performance assessment, will vary between each of the organisational scenarios outlined above. So too will the criticality of support service supply. Hence the key performance indicators used to assess the performance also need to be highly variable. Herein lies the nub of the performance assessment challenge for FM in the future.

13.10 Conclusion

The focus of contemporary FM performance assessment on the verifiable and quantifiable has limited the consideration of the wider, perhaps less tangible or differentiable value of FM. There is currently a divergence between the performance indicators preferred by the FM industry – which are tending towards generalisable, reductive databases for benchmarking the facilities-oriented aspect of FM – and those performance indicators which business is interested in, but which tend to be more synergistic and business-outcomes oriented. The consequential emphasis on measuring FM performance separate from the business has neglected its interactive value. This is a difficult issue to tackle, but one that ultimately lies at the heart of defining, managing, and assessing the contribution that requires assessment. A shift in emphasis will have to occur in the foreseeable future towards the assessment of the usefulness of FM; using business-oriented criteria for the quality of the business support service could help tackle this current limitation in assessing FM.

If, as appears likely, the future competitiveness of a significant proportion of the business world becomes increasingly pivotally dependent upon dynamic competitiveness rather than static competitiveness, then relative business advantage will be derived from the strategic application of FM in a customised manner, wherein the business support service is designed and assessed according to the specific strategic needs of a changing business. The indicators for assessing this aspect of FM performance will be have to be more high-level, more transparent to business, and will have to represent the synergistic business outcomes of a range of interrelated FM services, probably the result of a complex interactivity within and with the core business. They will also have to accommodate changing performance priorities as part of the performance assessment.

In these circumstances, the target business value of FM may reside in its usefulness when applied strategically for competitive business advantage. Hence the selection of key performance indicators for FM should be more directed towards business outcomes, and could beneficially converge with core business performance indicators such as agility, flexibility, business continuity, and/or transition management. This will allow the real value of FM to be demonstrated directly to business managers, and in terms that they can relate to. To achieve this, assessing the performance of FM must involve refocusing measurement from the outputs, processes, or simply the inputs of FM towards the business usefulness of its outcomes.

If, however, the current trend towards facilities-oriented generalisable and reductive data continues unattended, rather than supporting the evolution of FM into a strategic lever for business advantage, it would serve to normalise and hence obscure the potential of FM to contribute to business competitiveness on a bespoke basis.

However, the nearer FM comes to achieving the strategic usefulness and operational integration that the business *could* benefit from, the more impracticable it becomes to distinguish it from the core, and hence to measure its individual contribution to the overall outcomes. The remaining question over the continued evolution of FM as a whole is whether the paradox of measuring the value of an integrative service such as FM, which is as complex yet invisible as business process interactivity, and is then complicated further by its own operational and managerial processes, is simply too complex for FM or the business to be bothered about.

References

Becker, F. (1990) *The Total Workplace: FM and The Elastic Organisation*, Chap. 14, pp. 291–305. Van Nostrand Reinhold.

Chakravarthy, B.S. (1986) Measuring strategic performance. *Strategic Management Journal*, **7**, 437–58.

Dixon, J.R., Nanni, A.J. & Vollmann, T.E. (1990) The new performance challenge – measuring operations for world class competition. *Business One*, Irwin.

Fitzgerald, L., Johnstone, R., Brignall, S., Silvestro, R. & Voss, C. (1991) *Performance Measurement in Service Businesses*. CIMA.

Kaplan, R.S. & Norton, D.P. (1992) Using the balanced scorecard – measures that drive performance. *Harvard Business Review*, Jan.–Feb., 71–79.

Keegan, D.P., Eiler, R.G. & Jones, C.R. (1989) Are your performance measures obsolete? *Management Accounting*, June, 45–50.

Leaman, A. (1995) Dissatisfaction and office productivity. *Facilities*, **13**(2).

Massheder, K. & Finch, E. (1998) Benchmarking methodologies applied to UK facilties. *Management Facilities*, **16**(3–4), 99–106.

McDougall, G. (1999) *The aspects and purposes of facilities management performance measurement*. Unpublished paper, Heriot-Watt University.

O'Mara, C.E., Hyland, P.W. & Chapman, R.L. (1998) Performance measurement and strategic change. *Managing Service Quality*, **8**(3), 172–82.

Scott, M. (1998) *Value Drivers: The Manager's Framework for Identifying The Drivers of Corporate Value Creation*. Wiley.

Slater, S.F., Olsen, E.M. & Reddy, V.K. (1997) Strategy based performance measurement. *Business Horizons*, **40**(4), 37–44.

Tranfield, D. & Akhlaghi, F. (1995) Performance measures: relating facilities to business indicators. *Facilities*, **13**(3), March, 6–14.

Van de Vliet, A. (1997) Are they being served? *Management Today*, February, 66–69.

Varcoe, B. (1996) Business-driven facilities benchmarking. *Facilities*, **14**(3/4), March/April, 42–48.

Vorkurka, R. & Fleidner, G. (1995) Measuring operating performance: a specific case study. *Production and Inventory Management Journal*, **36**(1), 38–43.

Vorkurka, R.J. & Fliedner, G. (1998) The journey toward agility. *Industrial Management and Data Systems*, **98**(4), 165–71, MCB University Press.

Whitaker, M.J. (1995) Conducting a facility management audit. *Facilities*, **13**(6), June, 6–12.

14 Post-occupancy Evaluation

Danny Shiem-Shin Then

14.1 Introduction

The way buildings are designed and constructed continues to change in response to the evolving needs of the people and organisations that use them and as advances in building technology present new possibilities. We are living in an era of transformation – of technology, of social values, and of the way work is done (Aronoff & Kaplan, 1995). The goal of such transformation is to make the work of the organisation more efficient and productive – 'doing more with less' – i.e. to produce more with fewer resources and at a lower cost. In more recent times, the rapid growth in and accessibility to computing and communication tools is forcing corporate managers to reassess the basis upon which corporate decisions in facility provision and workplace management are evaluated.

In terms of business resources management, the real estate resource is generally accepted by corporate management as dealing with workplace (facility) provision and its ongoing management as a supporting business infrastructure needed to house the productive processes and tasks performed by its occupiers. In simple terms, buildings provide the workplace for the organisation. In a business environment for which the norm is change, it is critical that the real estate resource is managed efficiently and is effective in supporting the business outcomes by continuously adapting to align with the changing needs of the core business. It is in this context that the building's performance will be evaluated by the extent to which the facilities either support, or can be adapted to, the changing needs of the occupants it houses.

The growth of the facilities management market in the last two decades has given credence to the need by organisations to strategically plan their facilities or workplace requirements, both in form and quantity, and in timing of requirements. In many organisations, the facilities manager is charged with the role of workplace provision and management of operational support services within tight boundaries of service levels and occupancy costs limits. Whilst the technical complexity of facilities services operations (power, lighting, heating/cooling, etc.) and business support services (cleaning, call centre, reprographics, etc.) demand clear competencies, it should not be forgotten that facilities management is essentially a service business which implicitly means that 'the customer is king'. Herein lies the importance of post-occupancy evaluation as a tool for accessing the customer for consultation and valuable feedback, in addition to assessing their views on facilities solutions and services delivered.

There are numerous definitions of post-occupancy evaluation (or POE). The one proposed by Preiser *et al.* (1988) in their book of the same title is perhaps the most inclusive of key elements of the POE process:

> 'Post-occupancy evaluation is the process of evaluating buildings in a systematic and rigorous manner after they have been built and occupied for some time. POEs focus on building occupants and their needs, and thus they provide insights into the consequences of past design decisions and the resulting building performance. This knowledge forms a sound basis for creating better buildings in the future.'
>
> Preiser *et al.* (1988, p. 3)

This chapter will provide an overview of POE as a management investigative and consultative tool designed to evaluate building performance from an organisational context. The review will cover the use of POE in relation to building procurement and building performance, developments of POE techniques and research, the POE process models and POE applications.

14.2   The building procurement process and POE

The design, construction and occupancy of the built environment represent a complex process involving disparate groups of stakeholders from the producers' side (clients, project managers, financiers, architects, designers, contractors, regulating authorities) and the users' side (occupants, visitors, facilities managers, owners, maintenance personnel, special interest groups). The complex process highlights that the building life cycles are dynamic both organisationally and physically, involving a web of interacting social, economical, technical, functional and political systems (London, 1997). During occupancy, buildings undergo physical change as they age and deteriorate. At the same time, the occupants of buildings and their various organisations undergo functional and cultural change (Markus, 1972). Over the life cycle of a building, disparate groups of stakeholders collectively act to design, construct, manage and make decisions to create and alter buildings on behalf of others who occupy them. To fully understand its potential role, POE should be seen within the context of the building procurement process, in particular the critical relationship between briefing and the post-occupancy phases.

The conventional Plan of Work in the RIBA *Architect's Handbook* (RIBA, 1998) describes the process as a sequence of tasks from stage A (inception) to stage L (completion) and stage M (feedback). The feedback stage is often treated as an additional service following the hand-over of the completed building. Gray *et al.* (1994) also accepted that once the physical building has been completed then the consultants, contractors and subcontractors are no longer involved with the project. Law (1981) suggested a more formal feedback approach by showing how evaluation is inextricably linked with the briefing phase – the primary source of input into the subsequent design phases (Fig. 14.1).

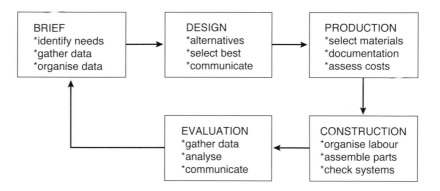

Fig. 14.1   Building process flow diagram.
Source: Law (1981, p. 8).

Research (Markus, 1967; Maver, 1970; O'Reilly, 1973; Preiser, 1989; Gray *et al.*, 1994; Latham, 1994) has largely supported the view that the briefing process in which buildings are designed and constructed is complex and highly fragmented both horizontally and vertically, and from phase to phase. Recent research (Bennett, 1990; McGeorge et al., 1994; London, 1996) indicates the importance of understanding the social complexity of client organisations and suggests that this may provide some answers to problems specifically in briefing and the flow on to evaluation.

14.3   Developments in POE techniques and research

The buildings we occupy today have evolved from a need for organisations to handle a greater volume of increasingly complex information. The office facility, in particular, has developed from a simple heated or cooled, lighted structure offering basic shelter from the elements, to a complex, climate-controlled, technology-intensive, and mission-critical information processing support infrastructure, typical of today's working environment.

There is growing support for the belief that the physical and social setting in which work takes place profoundly influences human performance (Becker & Steele, 1995). Organisations have become increasingly dependent for their success on the quality of work their employees deliver. Employee turnover, absenteeism, dissatisfaction, and premises-related illness and injury, all affect the level of productivity within the work environment and as a result compromise the effectiveness of the organisation's core capabilities. Yet in many organisations, decisions that are made that affect, in a fundamental way, the quality of the work environment are often in the hands of managers who do not have adequate knowledge of the needs of their operational facilities.

Occupiers who used them have always evaluated buildings. POE is a form of building evaluation that has evolved in response to a growing concern that completed buildings are not performing to the expectations of the occupants. In part it is a reaction to the deficiencies of architecture and the construction industry practices which focus mainly on the design and construction phases of the building life cycle, often to the exclusion of the eventual occupiers of the completed product during the building-in-use phase. In an attempt to systematically measure building performance and the extent to which the built environment did not satisfy requirements of stakeholders (i.e. owners, designers, builders and occupiers) in the building process, post occupancy evaluation (POE) emerged as a distinct area of research from the early 1960s. Preiser *et al.* (1988).

Initially, POEs were developed to evaluate building performance from a specific or narrow range of design or technical criteria. These evaluations were usually initiated to seek solutions to specific questions in relation to design aspects of the building, or the design and construction processes with a view to learning from mistakes made in order to improve process efficiency and effectiveness. The learning from such evaluations is problem-driven solutions.

The next phase of POEs in the 1970s widened the scope of building performance evaluation when it became clear that the dysfunctioning of the completed building could not be explained by considering only design and construction factors. Such evaluations must also incorporate the social context of the work environment that is embodied in the perceptions and opinions of the occupiers during the building-in-use phase of its life cycle. Concerns of building performance in such evaluations are primarily driven from the occupants' perspective with a focus on assessing occupant satisfaction in terms of environmental comfort parameters such as air-conditioning, lighting, acoustics and office layout. The energy crisis in the mid-1970s had the effect of narrowing the focus of building evaluations to energy conservation. This in turn led to a wave of occupant complaints about the environmental quality of offices – indoor air quality became another variable in building performance evaluations.

Apart from specialist technical skills, social science expertise is often required in the design of such evaluation forms. With the emergence of the PC and later the laptop computer, mobile telephone and internet communication, the scope of POE evolved to embrace the assessment of effectiveness of workplace reconfiguration by large corporations introducing alternative workplace strategies. These strategies are based on supporting an occupant's tasks by the provision of appropriate work settings rather than based on rank and status. From the mid-1980s, understanding the culture of an organisation became an important variable in POE initiatives aimed at assessing organisational effectiveness in relation to the performance of its reconfigured workplace provision.

Today, POE as a management tool is being integrated within the broader context of facilities management and strategic asset management. POE initiatives now focus on problems and issues that span the entire building delivery process. Its scope of evaluation has evolved from essentially a reality check of post-construction design features, to embracing multidimensional variables of a technical and social nature, covering a range of projects in office reconfiguration, corporate moves and relocation, introduction of new work practices, amongst others.

14.4 POE and building performance

The general push to performance management is driven by well-founded principles of effective and efficient use of business resources. More recently, the management drive has been for external comparative analysis as a basis for validating internal performance. The underlying rationale behind the push to measuring building performance stems from a realisation that the costs of construction represent only a fraction of the overall costs of ownership and that buildings are a means to various ends – they are the infrastructure support required for the delivery of products/ services. The acknowledgment that buildings, as physical assets, are durable products that require life-cycle management has created a need for better decision-making tools that are capable of evaluating alternative options in facility provision and management.

Over the last decade or so, there is a wider acceptance that the workplace environment has an impact on occupiers' productivity (Leaman & Bordass, 1997). The issue of user satisfaction in terms of fitness for purpose and quality of workplace environment has added the 'soft' human dimension to the technical and financial evaluation of performance of buildings in use. In a wider context within the property and facilities industry, there are indications that there is a coming together of facilities provision and facilities services management as a result of restructuring of the supply side of the market (Then, 2000). This trend is also evident from the maturing of the integrated service delivery market which offers an expanding service delivery scope that may include maintenance, cleaning and construction. On the part of corporate management there is also a closer scrutiny of facilities occupancy costs and a willingness to consider alternatives to in-house provision of a whole range of facilities support services.

The above brief review points to a wider management responsibility for asset and facilities managers with implications on how they will be assessed on the performance of the resource (the built asset base) under their care. One of the main challenges facing them will be to demonstrate how the assets under their control can add value to business processes, rather than being perceived as a continuing drain on scarce resources – or worse, a drain on profit. The development of any conceptual framework for evaluation of the performance of buildings (as operational assets) must recognise at least three important characteristics of buildings as a product and as a business resource:

(1) Buildings have a much longer life than most assets in business. A building represents a special class of durable assets requiring high initial capital investment and subsequent running costs and reinvestment. A regime of life-cycle management is required to optimise its efficient operation.
(2) A building's value is represented by its effectiveness as a supporting resource in the overall value chain of an organisation's productive process. Its role as an enabling resource is increasingly seen as crucial in raising staff productivity – an integrated resource management approach incorporating the delivery of the enabling workplace environment must be acknowledged.

(3) Buildings involve a number of owners, organisations, and users throughout their operational lives. Existing buildings are also being changed and renovated more often in response to new owners, organisational changes, and new occupant requirements – buildings as dynamic entities which must be managed proactively in order to respond to changing users' expectations and rapid technological development.

Evidence from the literature reviewed (Then, 1998) suggests that building performance monitoring is an amalgam of at least four aspects of facilities provision and their ongoing servicing as functional facilities:

(1) The appropriateness of the current asset base in meeting business objectives
(2) The provision of a satisfactory working environment for occupants and customers
(3) The minimisation of operating and maintenance costs due to the condition of the existing facilities
(4) The performance of the facilities as functional, operational assets

In optimising the performance of built assets, an organisation must balance the interdependent and, often competing, outcomes of the above four aspects of asset performance in order to achieve its optimum service potential. Taking the above constraints into consideration, Then and Tan (1988) proposed that building performance measures can be grouped under four broad categories or facets of asset performance measures:

(1) *Economic.* The economic facet of asset performance is principally concerned with decisions at a strategic level that optimises on value for money from property resources. Economic built asset management requirements are governed by the need to relate physical facilities provision to longer-term business plans. The objective of measurement here is to ensure optimum resource allocation and affordable and economic provision of property resources in line with market offerings and business plans.
(2) *Functional.* The functional facet of asset performance is principally concerned with management decisions that relate to the creation of the desired working environment in line with the preferred organisational culture and workplace standards. The objective of measurement here is to ensure continuous alignment of supply of appropriate functional space to anticipated

service demands as far as possible. Fitness for purpose for property resource in meeting business requirements may be measured in terms of type, form and size of buildings and distribution by location.

(3) *Physical.* The physical facet of asset performance is principally concerned with efficient and effective management of operational aspects of ongoing asset management. The objectives of measurement here are driven by the need to preserve asset value, ensure asset condition does not lead to unnecessary operational risks and liabilities, and to ensure occupancy costs are reasonable.

(4) *Service.* The service facet of asset performance is principally concerned with decisions and actions relating to quality perception by end-users and quality of service delivery by service providers. The objective of measurement here is to ensure that the business context and organisational culture are appropriately reflected in aspects of service delivery and are aligned with core business requirements.

The premise taken is that any integrated asset performance reporting must incorporate the four facets of measures in order to obtain a balanced view of the contribution of built assets as an operating resource, as shown in Fig. 14.2. It provides a basic structure for considering the many dimensions of building performance, and for critically reviewing the suitability of existing and potential measures.

Taking a more micro view, Preiser (1989) suggested a performance-based conceptual framework for considering the

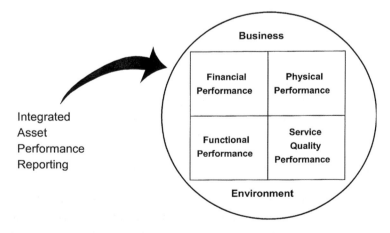

Fig. 14.2 Integrated asset performance reporting.
Source: Then *et al.* (2000).

interrelationship between the physical, functional and social variables in an organisational setting. The performance concept is based on the assumption that a building is designed and built to support, and enhance, the activities and goals of its occupants (Fig. 14.3). Preiser also suggested that improvements resulting from POEs can be grouped under three major categories: technical, functional and behavioural. The technical elements deal with survival issues, such as health, safety and security, and the performance of building systems. The behavioural elements deal with the perceptions and psychological needs for the building users and how these interact with the facility. The functional elements deal with the fit between the building and the clients' activities.

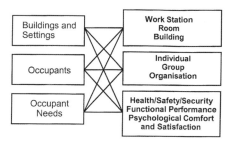

Fig. 14.3   Elements of building performance.
Source: Preiser (1989).

In POE the product being assessed is the functioning building, and the measurement is to assess how well the facilities are supporting the tasks and processes of the business. Given the number of building performance elements that potentially can be assessed, it is crucial that any POE initiative must be clear of its objectives and the motivation that drives the evaluation. The level of resources will dictate the amount of effort and specialist skills that are required to provide the necessary feedback for appropriate solutions to be formulated. In this respect the background research and choice of measurement methodology deserve serious consideration if the validity of the POE results is to be credible for appropriate action to be taken to effect improvements.

14.5   Developments in POE methodology

Post-occupancy evaluation is a form of building evaluation that has evolved and grown in status largely as organisations have sought to continuously align (or realign) mismatches between what is supplied in physical form (the building and its workspaces) and what is demanded in operational context (tasks and functions) in a dynamic business environment. POE is a systematic process of evaluating buildings after they have been built and occupied for some time. POEs focus on building occupants and their needs in a serviced environment. They provide insights into the consequences of past design decisions and the resulting building performance in meeting or not meeting set performance criteria. Access to this knowledge forms a sound basis for learning from past mistakes and as a result creating better buildings for the future.

Most books and articles have discussed POE as a set of techniques to collect information rather than as an approach towards better theoretical understanding of buildings (Zimring, 1989). More recently, POE evaluation research (Law, 1981; McGeorge *et al.*, 1994; London, 1996; Preiser *et al.*, 1988) has undergone a major shift, which accepts that for greater effectiveness the POE model must consider the entire building procurement life cycle.

Rabinowitz (1989) described the evolution of POE over the 20 years from the mid-1960s to the mid-1980s as progressing through three distinct periods in which POE became successively 'useful,' 'usable,' and 'used' and speculated that:

'The next breakthrough in POE will be in understanding larger, more dynamic environments with more choice and with the physical environment being only one of many factors influencing the performance of the people and the facilities they function.'

Rabinowitz (1989)

The latter breakthrough has already arrived as evidenced by recent literature supporting the crucial role of workplace environment as a catalyst in fostering a corporate culture of teamwork and continuous improvement and innovation. Becker and Steel (1995), McGregor and Then (1999).

Table 14.1 summarises the key development characteristics of POE. POE has developed from its early applications in the 1960s, where the majority of formal POE processes were developed and used by public sector bodies of various types and large corporate organisations, evolving in recent years to a valuable management tool for measuring and tailoring operational building performance to meet changing user needs. Early initiatives of POE in North America (1960s and 1970s) have been linked with public sector organisations involved in repetitive building programmes where the focus was on learning from previous design failures as the basis for ongoing, iterative improvements in facilities and enhancement of quality (Becker & Steele, 1995; Vischer, 1996; McGregor & Then, 1999). The 1970s saw the rapid expansion of POE activity in terms of

Table 14.1   Development of POE methodology and its growing acceptance.

Development phase	Developmental characteristics

1960:
Early initiatives
Pioneering studies
	Useful results from early POEs:
	  Formal and comprehensive examinations of
	  buildings were found to have benefits for users and
	  owners of buildings, as well as architects designing
	  similar buildings.
	Key learning: Learn from previous mistakes.

1970s:
Gaining acceptance
Systematic POEs
	Developing usable methods:
	  Systematic processes and methods that are reliable
	  and replicable promote accessibility and use by a
	  larger audience. Encourages applied research and
	  interdisciplinary approaches.
	Key learning: Rigorous methodology gains credibility.

1980s:
Widening scope
Applied POEs
	Becoming routinely used:
	  The benefits of POE acknowledged; POE process
	  become widely known and frequently used and
	  commercially accepted. Specialisation in POE
	  introduced in curricula of graduate programmes in
	  architecture.
	Key learning: Investments in POE must prove its value.

1990s:
Recognition of facility–
productivity link
Professional POE
services
	Incorporated in building delivery process best practice
	guidelines/manuals:
	  POE accepted as part of tools essential for effective
	  building delivery. Growth of facilities management
	  reinforces the crucial link between workplace
	  environment and occupiers' productivity.
	Key learning: The workplace influences behaviour –
	staff productivity a major management concern.

Source: Adapted from Rabinowitz (1989).

knowledge of evaluation methods, building types and occupant groups (Preiser & Daish, 1983). The widening scope of POE was matched with increased methodological sophistication with the inclusion of a social-science-based approach in designed evaluation with a focus on user satisfaction covering various aspects such as setting, clients, environmental context, design process and social-historical context. The 1980s saw the acceptance of POE as a discipline that warranted inclusion in graduate training in architecture and facilities planning and management. This was a period when the process methodologies of POE's practices were consolidated and formalised and gained gradual inclusion in the overall briefing–design–construction–occupancy building process. The rapid pace of development in information technology during the late 1980s and 1990s and its impact on work and the workplace, has had the effect of bringing about a greater awareness by organisations of the need to align the physical workplace with the tasks that people perform in adding value to the business they are supporting. POE services have evolved to meet demand for evaluations that assess the effectiveness of office reconfiguration and moves that are instigated by efforts to change organisation culture, introduce new technology, and promote continuous improvement. POE's continued relevance will be influenced, to a large extent, by its ability to incorporate clients' business requirements and performance criteria in its design and execution.

14.6   POE process models

POE is primarily concerned with evaluating buildings in use. The value of POE lies in its ability to assist in the quality control of the building and the building process. Both the scope and coverage of POE have continued to evolve. White (1989) noted that advances in POE methodologies have led to a gradual shift in the definition of POE from the 'evaluation of buildings' to 'service to the client' in a more holistic sense. This implies that both technical and organisational (contextual) criteria should be used in determining the overall success of POE studies. Figure 14.4 illustrates some of the technical and contextual influences on the scoping of POE objectives and factors that impact on the planning of POE studies. As mentioned earlier, the POE process is underpinned by the performance concept as it is applied to the building process.

Conceptually, the context of POEs can be considered within a performance evaluation research framework (Preiser *et al.*, 1988). The framework (Fig. 14.5) connects the evaluation of buildings with three important elements:

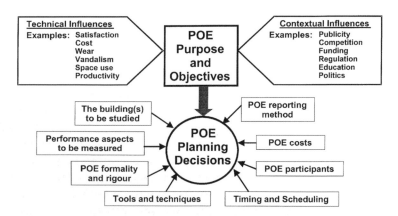

Fig. 14.4   POE objectives and planning decisions.

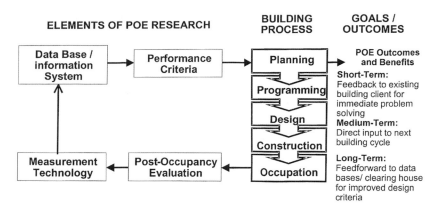

Fig. 14.5   The performance evaluation research framework.
Source: Adapted from Preiser *et al.* (1988).

(1) *Measurement technology.* This includes techniques and techno-
      logical aids that are used in the data collection and analysis of
      POE. They include interviews, questionnaire surveys, direct
      observation, mechanical recording of human behaviour, mea-
      surement of light and acoustic levels, recording with video and
      camera. Standardisation of measurement technology is an issue
      that impacts on comparison of results with other studies.

(2) *Databases and information systems.* Data collected from appro-
      priate measurement technology can be fed into databases,
      information systems, or clearing houses that contain the results
      of other POEs. Design guidelines are often distilled from such
      sources.

(3) *Development of performance criteria for buildings.* Performance
      criteria and guidelines can be developed from databases and
      other information systems for a given organisation and/or
      building type, usually documented in either technical manuals,
      design guides, or specialised databases.

Figure 14.5 illustrates the potential benefits to be derived from
properly organised and collated POEs in the short, medium and
long term. Apart from the identification and solutions to immediate
problems in facilities, valuable insight gained from the evaluation
can be feedback to improve the building process. In the longer term,
the accumulated knowledge from recorded POEs can be distilled to
improve design and performance criteria in operational and design
guides. While there are variations in the motivation to initiate
building-related studies and the way POEs have been carried out by
organisations, researchers and consultants, a generic model for the
POE process has been suggested by Preiser *et al.* (1988) in their book.

Their research and practice experience led them to suggest a three-phase process which is categorised according to the level of effort required to conduct the POE based on the amount of time, resources and personnel; the depth and breadth of evaluation; and the cost involved. These three levels of POE – indicative, investigative and diagnostic – are distinct and not cumulative.

- *Level 1. An indicative POE* simply provides an indication of major failures and successes of a building's performance. It is usually carried out within a very short time span, from a few hours to one or two days. Performance criteria chosen are based on the experience of the evaluator. Typical data-gathering methods include archival and document evaluation, walk-through evaluation, evaluation questionnaire survey and selected interviews.
- *Level 2. An investigative POE* is more time-consuming and complicated and more resource-intensive. Investigative POE uses researched criteria derived from state-of-the-art literature assessment and recent comparison data that are objectively and explicitly stated. Investigative POEs are often a follow-up when indicative POEs have identified major issues that warrant more detailed study.
- *Level 3. A diagnostic POE* is a comprehensive and in-depth investigation conducted at a high level of effort utilising scientific methodology that is very similar to traditional, focused research. The evaluation may take from several months to a year, or longer. Its results and recommendations are long-term oriented and focus on a given type of facility or building.

A more procedural-oriented POE process model is given by the Queensland Department of Public Works in its *Strategic Asset Management Guidelines* (Queensland Government, 1997). Part 7.5 of the Guidelines describes a POE as a structured approach for the evaluation of the performance of a new or improved facility against asset planning requirements, once fully operational (after 12 months). The POE is viewed as a three-phase process designed to the collection and analysis of data in relation to facility performance and the translation of these findings into action plans (Fig. 14.6). It advocates POE as one of a suite of resource planning reviews that may be applied to projects within the context of strategic asset management.

At a more conceptual level, McGeorge *et al.* (1990) proposed a dynamic systems model in which POE forms part of an integrated decision support framework with stakeholder representation from both users' and producers' groups over the project life cycle of inception–design–construction–occupancy. They proposed that

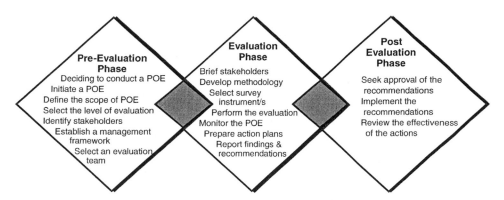

Fig. 14.6   The post-occupancy evaluation process.
Source: Queensland Government (1997).

POEs can be more proactive by contributing to producers' decision-making by not only providing feed-forward information but can also trigger and highlight issues that will require input from users.

14.7   Applications of POE

POE has largely dealt with the relationship between the building and the user of the building within a social and cultural environment. This focus will largely remain. However, the emergence of the facilities management field in the late 1980s and 1990s saw a broadening of the scope of evaluation covered by POE projects to include monitoring of customer service elements of building in use. This section will provide some examples of POE projects spanning the traditional one-off technical evaluation of buildings to the provision of periodic feedback to clients. The feedback can take the form of reporting on the impact and effectiveness of changes within the organisation as a result of policy changes relating to workplace configuration; introduction of new technology and work practices; or facilities services delivery monitoring. The source references give a more detailed explanation of the process and outcomes. The examples are:

(1) Recurring construction programmes: public sector housing, hospitals, universities, correctional facilities – learning from feedback of recurring programmes.
(2) Technical appraisal of design solution of new/existing buildings – feedback of design criteria from users, for example:
   - *POE for the Visual Arts and Science Buildings* (1999), unpublished internal report, Abbato, Office of Facilities Manage-

ment, Gold Coast Campus, Griffith University, Queensland. The survey measured the levels of satisfaction for students, academics, general staff and building operator who use the visual arts buildings. Issues covered in the survey were grouped into seven categories: achieving university and educational objectives; the building generally; site planning and access; the design of individual rooms and facilities; internal aspects; human comfort; and comments and suggestions.

- *A Study of Post Occupancy Evaluation within University Student Residential Accommodation* (1999), Blair, C, unpublished MSc dissertation, Heriot-Watt University, Edinburgh. The survey covered the evaluation of four aspects of university student residential accommodation in the UK: technical, functional, behavioural and statutory compliance factors.

(3) Technical appraisal to user focus survey of facilities services:
- *Occupancy Survey for the Asian Studies and Technology Learning Centres* (2000), unpublished internal report, Abbato, Nathan Campus, Griffith University, Queensland.

(4) Periodic survey of user satisfaction of facilities and services:
- Powerlink, Queensland – periodic users' feedback monitoring of facilities performance following the introduction of a new way of working in a remodelled facility aimed at changing organisational culture using the working environment as a catalyst for change (McGregor & Then, 1999).

(5) Programming and design consultation:
- POE of facilities – a participatory approach to programming and design (see Horgen & Sheridan, 1996). Case study of Kennedy School of Government, Harvard University.

(6) Feedback from evaluation of design–construction–occupancy process:
- UK PROBE studies (Chartered Institute of Building Services Engineers – CIBSE Building Services July 1995). Case studies of buildings with the aim of improving understanding of how buildings and their services are being operated, controlled and maintained. The range of post-occupancy issues assessed includes:
  ○ Design and construction
  ○ Design integration
  ○ Effectiveness of the procurement
  ○ Methods of construction, installation and setting to work
  ○ Initial occupation of the building

(7) Japanese POE studies. The Joint Study of Government and Private Companies on Developing a Standard Post Occupancy Evaluation for Office Buildings:

- A series of papers presented at the 1990 CIBW70 Tokyo Symposium (in Proceedings of CIBW70 Tokyo Symposium – Strategies and Technologies for Maintenance and Modernisation of Buildings, 253–308).

From the selective review of POE applications, it is clear that the use of POE must be driven by clear objectives commensurate with the amount of resources available. POE as a management tool, when carefully organised, provides access to tap valuable feedback from users of facilities. Learning from feedback is a key motivation of the POE process. Effective communication is a crucial ingredient of any POE studies. As the value of POEs becomes increasingly accepted and acknowledged by client organisations, it will have a role to play as a decision-making tool for organisations to manage their built (operational) assets actively and carefully (Zeisel, 1981; Shibley, 1985; Preiser *et al.*, 1988).

14.8 Conclusion

Ultimately, the effective management of buildings is about the fit between the facility and its users. POE as a tool has evolved to measure this match with the aim of gaining a better understanding between form and functions, workspace and tasks, organisational culture and working environment. The growing acceptance of POE has been supported by tools and techniques that attempt to bridge the divide between the design, construction and building-in-use phases of the building life cycle, both in terms of communication during project brief formulation and construction period, as well as feedback from the occupiers of the building in use.

In balancing the need for technical competence in research rigour and a successful outcome from the client's perspective, it is important to remember that buildings as a business-operating resource are long-term capital assets which possess unique attributes that must be recognised in their ongoing management:

- Buildings have a much longer life than most other assets in business.
- Buildings involve a number of owners, organisations, and users throughout their lives.
- Existing buildings are being changed and renovated more often in response to new owners, organisational changes, and new occupants' requirements.

In the main, POE initiatives provide a critical and strategic understanding of the workspace requirements necessary to bring about a closer alignment between the physical settings provided by the building form, the organisation culture defining the functions, and the way users of the facility are to be supported. POE is well-placed to contribute within this wider business resources management framework. However, for this to happen the 'producers' and 'users' (London, 1997) of the building process must be more integrated – activities of briefing, designing, constructing, post-occupancy evaluation and refurbishment must be seen as part of an ongoing system within a large social environment.

References

Abbato, N. (1999) *Post Occupancy Evaluation for the Visual Arts and Science Buildings – Griffith University*. Internal document from Office of Facilities Management, Griffith University.

Abbato, N. (2000) *Occupancy Survey for the Asian Studies and Technology Learning Centres – Nathan Campus, Griffith University*. Internal document from Office of Facilities Management, Griffiths University.

Aronoff, S. & Kaplan, A. (1995) *Total Workplace Performance – Rethinking the Office Environment*. WDL Publications.

Becker, F. & Steele, F. (1995) *Workplace by Design – Mapping the High Performance Workscape*. Jossey-Bass Publisher.

Bennet, J.F.D. (1990) Specialist contractors: a review of issues raised by their role in building. *Construction Management and Economics*, **12**, 551–6.

CIBSE (1995) Post-occupancy review of building engineering. *Building Services*, July, 14–16.

Gray, C., Hughes, W. & Bennett, J. (1994) *The Successful Management of Design*. Centre for Strategic Studies in Construction.

Horgen, T. & Sheridan, S. (1996) Post-occupancy evaluation: a participatory approach to programming and design. *Facilities*, **14**(7/8), 16–25.

Latham, M. (1994) *Constructing the Team* (the Latham Report). Department of the Environment, Transport and the Regions, HMSO.

Law, N.M. (1981) *Evaluation of the post-occupancy performance of buildings – a state of the art report*. Experimental Building Station (EBS), Department of Housing and Construction.

Leaman, A. & Bordass, W. (1997) Productivity in buildings: the killer variables. Paper given at *Workplace Comfort Forum*, London, October 29–30.

London, K. (1996) Paradigm poised: charting the post occupancy evaluation. Paper given at *30th Conference of the Australia and New Zealand Architectural Science Association*, Hong Kong, Chinese University.

London, K. (1997) *The development of a post occupancy evaluation model based on a systems approach*. Master of Building thesis, University of Melbourne, pp. 6–7.

Markus, T.A. (1967) *The Role of Building Performance Measurement and Appraisal in Design Method*. The Architect's Journal Information Library (20 December).

Markus, T.A. (1972) *Building Performance*. Applied Science Publishers Ltd.

Maver, T.W. (1970) *Emerging Methods in Environmental Design and Planning Appraisal in the Building Design Process* (ed. G.T. Moore). MIT Press.

McGeorge, W.D., Chen, S.E. & London, K.A. (1990) The use of post occupancy evaluation in developing a model of buildability. Paper given at the *CIB W70 Tokyo Symposium*, pp. 899–906.

McGeorge, W.D., Chen, S.E. & London, K.A. (1994) The use of post occupancy evaluation in developing a model of buildability. Stategies and technologies for maintenance and modernisation of building. Paper given at the *CIB W70 Tokyo Symposium*, International Council for Building Research Studies and Documentation, CIB, **2**, 899–906.

McGregor, W. & Then, S.S. (1999) *Facilities Management and the Business of Space*, pp. 129. Arnold.

O'Reilly, J.J.N. (1973) *A case study of a Design Commission*. Building Research Establishment.

Preiser, W.F. (1989) *Building Evaluation*. Plenum Press.

Preiser, W.F.E. & Daish, J. (1983) *Post-Occupancy Evaluation – A Selected Bibliography*. No. A-896, Vance Bibliography.

Preiser, W.F.E., Rabinowitz, H.Z. & White, E.T. (1988) *Post-Occupancy Evaluation*. Van Nostrand Reinhold.

Queensland Government (1997) Part 7.5 Post occupancy evaluation. In: *Strategic Asset Management Guidelines*. Building Division, Department of Public Works and Housing.

Rabinowitz, H.Z. (1989) The uses and boundary of post-occupancy evaluations: an overview. In: *Building Evaluation* (ed. W.F.E. Preiser). Plenum Publishing Corporation, pp. 9–18.

Royal Institute of British Architects (1998) *Architect's Handbook of Practice Management*, 6th edn. RIBA Publications.

Shibley, R. (1985) Building evaluation in the main stream. *Environment and Behavior*, **17**(1), 7–23.

Then, S.S. (1998) *A framework for assessing building performance in the public sector*. Unpublished internal report to Department of Public Works, Queensland.

Then, S.S. (2000) The role of real estate assets in supporting the fulfilment of corporate business plans – key organisational variables for an integrated resource management framework. *Facilities*, **18**(7/8).

Then, S.S. & Tan, T.H. (1988) A performance measures framework for asset management in Queensland. Paper given at *CIB W70 Singapore '98 Symposium on Management, Maintenance and Modernisation of Building Facilities – The Way Ahead into the Millennium* (ed. Q.L. Kiang), 18–20 November, pp. 87–94.

Then, S.S., Tan, T.H. & Barton, R. (2000) 'Public Sector Asset Performance – Concepts and Implementation', *Proceedings of CIB W70 Brisbane International Symposium on Providing Facilities Solutions to Business Challenges* (ed. Danny S.S. Then), November 2000, pp. 17–30.

Vischer, J.C. (1996) *Workspace Strategies – Environment as a Tool for Work*. Chapman & Hall.

White, E.W. (1989) Post-occupancy evaluation from the client's perspective. In: *Building Evaluation* (ed. W.F.E. Preiser). Plenum Publishing Corporation.

Zeisel, J. (1981) *Inquiry by Design: Tools for Environment–Behaviour Research*. Brooks-Cole.

Zimring, C.M. (1989) Post-occupancy evaluation and implicit theory: an overview. In: *Building Evaluation* (ed. W.F.E. Preiser), pp. 113–25. Plenum Publishing Corporation.

15 Sustainable Building Maintenance

Keith Jones

15.1 Introduction

For many years building maintenance and refurbishment has been regarded by many as one of the most unattractive aspects of the building process. It has been seen as a 'low-tech, dirty' activity that belongs in the back room, rather than the boardroom. This perception significantly undervalues an activity that accounted for approximately £30 billion pounds (50% of construction activity) in the UK in 1999 (DETR, 2000a). This chapter challenges this perception by summarising current best practice in built asset maintenance and refurbishment (BAM&R) and highlighting how the adoption of such practices places BAM&R in the strategic resource management framework. The summary can be divided into three topics:

(1) A review of the theory underpinning maintenance and refurbishment
(2) The development of a maintenance and refurbishment strategy
(3) The development and execution of maintenance plans

The final section of this chapter examines the impact that the sustainability agenda may have on BAM&R practice and highlights the challenges that following the sustainability agenda may place on current best practice and on the fundamental logic that underpins contemporary thinking. For this final section of the chapter, reference is made to UK Government initiatives, and research activities with an international panel of maintenance and refurbishment practitioners.

15.2 The theoretical basis for maintenance and refurbishment

Before examining the theory underpinning maintenance and refurbishment, it would be useful to define the limits of what is being discussed. To this end most maintenance texts take the British Standard definition of maintenance (BSI, 1966) 'a combination of any actions required to retain an item in, or restore it to, an acceptable condition' as their start point. The limitations of this definition are graphically illustrated in Fig. 15.1 in which Finch (quoted in Alexander, 1997) draws a distinction between building maintenance and facilities management. In this diagram the building's capabilities (value) are plotted against operation (time). From the point of inception the building's performance starts to deteriorate. This deterioration can be considered as a combination of three separate components: physical decay, increased functional demands and technology improvements.

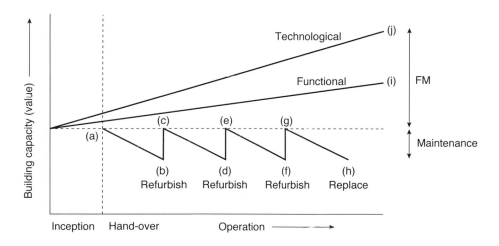

Fig. 15.1   Distinction between building maintenance and facilities management.
Source: Alexander (1997).

At hand-over (a) the physical condition of the building starts to deteriorate. At some point in the future (b) the degradation in the building's capabilities reaches a point at which a decision is made to refurbish the fabric to return it to a status quo condition (c). (Note: this may be the original condition or represent a slight improvement on the original condition.) This process of decay and refurbishment is repeated on a cyclical basis (d–h) until a point is reached where the performance of the building fails to satisfy the occupiers' demands. This performance gap may be due to changes in working practices (i) or the application of new technologies, particularly information technology (j). Thus, if maintenance and refurbishment is considered solely as a building technology issue then, no matter how diligently the building is maintained, it will inevitably become functionally and technically obsolete over time.

In recognition of the existence of obsolescence, the CIOB (1990) proposed an alternative definition of maintenance and refurbishment: 'work undertaken in order to keep, restore, or improve every facility, its services and surrounds to a currently acceptable standard and to sustain the utility and value of the facility.' In this definition we see an acceptance that maintenance and refurbishment is linked to improving the value of the built assets. Taking this definition as your start point, it is possible to reinterpret Finch's (quoted in Alexander, 1997) model to one which probably more accurately reflects what happens in practice. (Note: The latter model was developed from discussions between the author and Mr C. Burrows, BSc, FRICS, MCIOB, FBEng.)

The first part of this refined model (Fig. 15.2) is similar to that proposed by Finch. At hand-over (a) the physical condition of the building starts to deteriorate and routine maintenance begins. At some point in the future (b) the degradation in the building's capabilities reaches a point at which a decision is made to repair the fabric to return it to the status quo condition (c). This process of decay, routine maintenance and repair is repeated on a cyclical basis (a–d) until a point is reached where, even following routine maintenance and repair (d–e), the performance of the building fails to satisfy the occupiers' demands (e–f). At this point a major refurbishment (e.g. a technical upgrade of internal services, redesign of internal space layouts, etc.) of the built asset is required (d–f). Note: even a major refurbishment will be unlikely to address all the changing demands placed on a building and some residual obsolescence, caused by factors that are outside the control of the

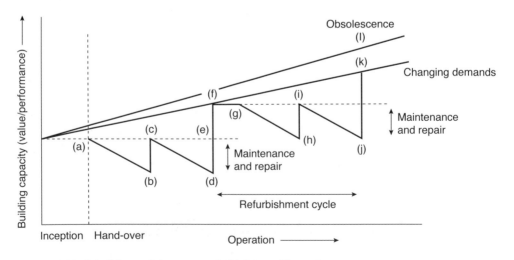

Fig. 15.2 Model of the maintenance–refurbishment life cycle.
Source: Adapted from Finch (quoted in Alexander, 1997).

building managers (e.g. economic influences, financial restraints, etc.), will remain. The decay, routine maintenance, repair and refurbishment cycle is now repeated (f–j) until some point in the future when, even with a major refurbishment, the obsolescence gap (k–l) is too great for the organisation to bear. At this point the organisation will need to either relocate or demolish and rebuild.

Although the model presented here is an obvious simplification of reality (buildings do not behave as single entities but as complex mechanisms), it does provide a good focus from which to view the maintenance, repair and refurbishment cycle. In particular, the model draws attention to a number of key decision points:

- What is the role of routine maintenance and repair in the context of the value (performance) of the built asset?
- What level of deterioration in value (or performance) of the built asset can be tolerated before routine maintenance, repair or refurbishment must take place?
- What impact does the obsolescence gap have on the performance of the organisation?

Consideration of these issues takes the maintenance and refurbishment of built assets beyond merely a building technology issue to one that acknowledges their impact on the long-term future of an organisation. Once this is accepted, then built assets become a strategic resource that has to be managed within the broader context of the organisation's strategic planning. Indeed, the positioning of BAM&R within an organisation's management framework attracted considerable debate during the 1990s. The outcomes of this debate were summarised by Quah (1999) who presented an overview of the relationship between facilities management and building maintenance (Fig. 15.3). In her discussion, Quah drew attention to the changing perception of BAM&R from one of a liability and nuisance, to one of supporting the core income generation activities of an organisation. Once you consider BAM&R as contributing directly to the profitability of an organisation then you elevate it from a backroom (technical) activity to one of strategic importance that has to be managed. This poses the first of the challenges alluded to at the beginning of this chapter – namely, the way in which those professionals involved in maintenance and refurbishment present their strategic arguments.

15.3 Maintenance and refurbishment strategy

All organisations, no matter how small, will have some form of corporate strategy. This corporate strategy provides the organisa-

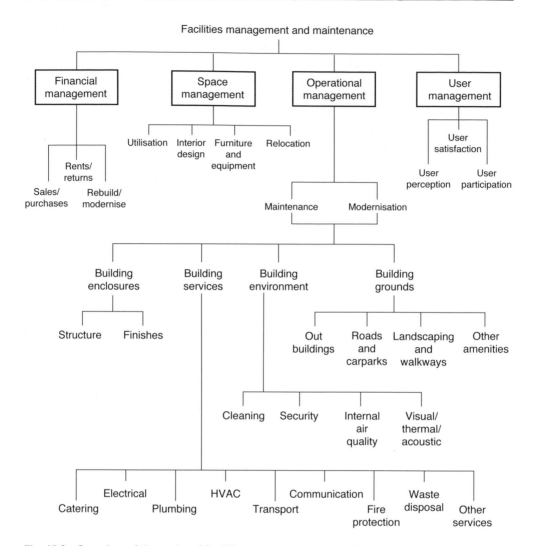

Fig. 15.3   Overview of the realm of facilities management and maintenance.
Source: Quah (1999).

tion with a sense of direction and specifies its objectives, purposes and goals. The corporate strategy also identifies the policies and plans that are necessary to achieve these goals. As the strategic direction of the organisation will determine its built asset requirements, and in turn, the built assets' performance will affect the organisation's ability to achieve its objectives, the role of maintenance and refurbishment must form part of the policies and plans that constitute the corporate strategy. Thus maintenance and refurbishment must be considered as a strategic activity. In its simplest form the maintenance and refurbishment strategy presents a clearly argued set of principles that show how an organisation intends to develop its built assets over the next 5, 10 or 20 years (depending on the organisation's time frame). At the detailed level, the maintenance and refurbishment strategy consists of a series of explicative policy and planning documents that identify the objectives and describe processes that the organisation intends to adopt to keep its built assets in a state fit for purpose. In general terms the strategy should identify:

- The life expectancy of the built asset (and of its major components)
- Any anticipated changes in building use or occupancy
- The resource base necessary to ensure compliance
- The management structure within which the maintenance and refurbishment process occurs.

However, the maintenance and refurbishment strategy does not take the form of operational documents; they are management documents, written for, and used by, people who are not building experts. As such, they should be written without recourse to technical jargon. This is not to imply that they are simplistic documents; on the contrary, they must present complex arguments in a clear and concise manner that informs management decision-making. Only then will BAM&R be taken seriously at the senior levels of an organisation. The following sections outline the general issues that the maintenance and refurbishment strategy would normally address.

15.4   The role of maintenance and refurbishment within an organisation

It is essential that the objectives of the BAM&R strategy be considered in the context of an organisation's future business plan. For example, regular building disposal and acquisition, probably on a lease basis, may better serve the business goals of an organisation and therefore the objectives of the BAM&R strategy would be to do the minimum necessary to satisfy any lease conditions (probably phases a–c of Fig. 15.2). Alternatively, if the built asset represents a significant investment to an organisation, the objectives of the BAM&R strategy would be to ensure that they retain, or if possible improve, their value over time. In considering their perspective, an organisation should critically assess the current/future needs of their built assets in terms of, for example, location, size, layout and performance standards and identify any inadequacies. The organisation is then able to clearly articulate its reasons for occupying its built assets, determining whether they are:

- Perceived primarily as an investment or an income stream.
- Fundamental to the primary business of the organisation (e.g. a Housing Association).
- Solely a necessary overhead to allow the primary function of the organisation to continue (e.g. a factory).

In seeking solutions, the organisation can consider outline possible/ acceptable solutions, for example refurbishment, acquisition, disposal.

When considering each of the above, an organisation must pay due regard to any tenancy and legal constraints that apply to their built assets. These may include, for example:

- Lease terms and conditions
- Health and safety regulations; fire safety regulations
- Building regulations
- Town and country planning regulations
- Party wall legislation
- Shops, Offices and Railways Premises Act; Landlord and Tenant Act
- Disability Discrimination Act, etc.

Once all of the above have been considered then the objectives of the BAM&R strategy can be articulated as a series of explicative statements that describe an organisation's perspective on their built assets.

15.5 Performance measures for buildings in use

All built assets deteriorate over time in respect of their physical, functional and technological performance. It is important that an organisation understands this fact and that criteria exist against which the deterioration can be measured and the outcome of maintenance and refurbishment actions judged. Thus the BAM&R strategy should contain explicit performance criteria that will allow future maintenance and refurbishment plans to be developed. Performance criteria should be developed for: cleaning the interior and exterior; the performance of the building fabric (both functional and decorative); the building services both in terms of performance and environment, e.g. an extractor unit make be capable of removing the required amount of air but may be producing an unacceptable level of noise pollution. When establishing the criteria, due consideration should be given to, for example, the cost implications, the methods of objective assessment wherever this is a possible, and statutory obligations.

The second group of performance measures that need to be addressed in the BAM&R strategy relate to the priority-setting criteria that will be used to programme maintenance and refurbishment activity. In its simplest form this involves making a judgement on the urgency of an activity. In this section of the BAM&R strategy, the philosophical basis (either quantitative and/or qualitative assessments) upon which priority decisions will be made and the scales of measurement that will be adopted should be outlined. When establishing priority criteria, due account should be taken of the following:

- Building status – the relative importance of the building to the organisation.
- Physical condition – the likelihood of complete failure of the element and its implication on other parts of the building.
- Importance of the element – whether the non-functioning of the element can be accommodated.
- Effect on users – health and safety issues must be prioritised above aesthetic concerns.
- Effect on primary business services – cost implications of loss of primary business services due to complete failure.
- Statutory obligations – these normally place time constraints on activities.

The final consideration when establishing performance measures is organisational politics. Those individuals who are responsible for the organisation's overall business strategy will examine the BAM&R strategy. Whilst they may not be building technologists, and as such may not be interested in the subtleties of maintenance and refurbishment practice, they will be interested in the implications of the strategy to their particular area of the organisation. Thus they will pay particular attention to this section of the BAM&R strategy as the decisions reached on performance criteria and priority-setting will directly affect their working environment. During the discussion period, various power groups within an organisation will attempt to ensure that the approaches proposed in the BAM&R strategy do not adversely affect their working environment. As such it is essential that the strategy contains sufficient information to allay their concerns whilst not restricting the maintenance manager when hard decisions have to be made in the future.

15.6 Approaches to maintenance and refurbishment works

At some stage in the life cycle of an organisation the BAM&R strategy will have to be translated into a series of operational maintenance plans. This process can be significantly simplified if, during the development of the BAM&R strategy, the alternative maintenance and refurbishment options available to an organisation have been evaluated. Generally speaking, maintenance can be dived into those actions that are preplanned and those that occur in an unplanned manner (BSI, 1966). Planned maintenance can be further broken down into preventative maintenance or corrective maintenance. The aim of preventative maintenance is to reduce the likelihood of failure or performance degradation of a building component. Preventative maintenance is either carried out according to a predetermined schedule or in response to an assessment of a component's condition. Corrective maintenance, under a planned maintenance system, occurs when it is apparent that a component will fail and the failure is planned for although it is not known exactly when it will fail. Unplanned maintenance, as its name suggests, covers those activities that are not preplanned and may take the form of corrective or opportunistic maintenance actions. In reality, very few maintenance activities occur in a totally unplanned manner. Even actions required after an unforeseen disaster, for example a flood, should be considered to some degree in a disaster recovery plan.

The BAM&R strategy should identify which approach to maintenance an organisation intends to adopt, for example on a component-by-component basis. In establishing the appropriate approach, account should be taken of the:

- Failure mode of the component (e.g. health and safety)
- Characteristics of the material (e.g. implications on surrounding components)
- Cost to the organisation (e.g. loss of core business function)
- Impact on the user (e.g. inconvenience)

For each of the general approaches, appropriate repair strategies should be evaluated. Repair strategies include:

- Operation to failure – in which the component is used until it fails, at which point it is replaced.
- Condition assessed – in which maintenance is undertaken when a component falls below a predetermined condition.
- Time period – in which maintenance activities are undertaken according to a predetermined time-scale.

- Opportunistic – in which the opportunity is taken to maintain one component whilst working on another.
- Design out – in which components are upgraded as a consequence of maintenance rather than being simply returned to their original condition.

Again, in making these decisions, regard must be paid to the cost implications, including the implications of failure to the core business, any statutory obligations and to the priority criteria assigned to the built asset component.

15.7   Procuring maintenance and refurbishment works

There are many different ways of procuring and managing building maintenance and refurbishment. The BAM&R strategy should outline the mix of approaches that an organisation intends to adopt. Whilst the exact make-up of approaches will largely depend on the characteristics of the organisation, at the strategic level, two overriding issues have to be considered:

(1)   The source of the activity (in-house or outsourced)
(2)   Contractual arrangements (between departments or organisations)

In general, maintenance activities can be divided between those performed by in-house staff and those which are put out to contract. This division applies equally to maintenance management functions (outsourcing) and maintenance operations (contracting out). At the strategic level, an organisation has to consider the implications of both approaches to its core business.

While it is currently unclear as to the long-term impact of outsourcing on an organisation (see Table 15.1), it is clear that many believe it to be the way of the future. This view may be further re-enforced if the principles of strategic partnering are applied to maintenance and refurbishment activities, as they are currently being to new build. In this case the adoption of long-term relationships between client organisations and maintenance service providers could well tip the balance in favour of maintenance management outsourcing.

Whilst there is still considerable debate over the outsourcing of maintenance management, most organisations already contract out their maintenance and repair activities, with in-house maintenance operatives being retained only where highly specialist work exists. In general, maintenance work is contracted out either on a project-

Table 15.1   Advantages and disadvantages of outsourcing.

Advantages	Disadvantages
Allows management to concentrate on core business activities	The waste of management time dealing with contracts Poor control by both parties and ambiguous accountability
Facilitates downsizing during a period of business reorganisation	Less control of services and increased dependence on another organisation's procedures for service levels
Provides greater flexibility in numbers and skill base of operatives	A loss of in-house expertise potentially removing the option of bringing the work back in-house later
Encourages economies of scale for the contractor	Extra costs being incurred to monitor service delivery
Improves control of budgets	A greater risk of being locked into commercial contracts that subsequently lose relevance
Provides access to breadth of specialists and equipment	A loss of knowledge of building and organisational practices and procedures Contracts not being as precise and detailed as they should have been

by-project basis or as part of a term contract. Project-based contracts cover a particular job. Term contracts are similar in concept to partnering arrangements in which the organisation agrees to place a proportion of its maintenance work with a contractor for a fixed period of time. Term contracts may take a number of forms depending upon the type of work being undertaken.

- Measured term contracts consist of a detailed schedule of priced work activities that form the basis of a bidding document against which contractors tender an adjustment percentage. Once the contract has been agreed, any work carried out by the contractor is measured on completion and the contractor reimbursed in accordance with the value set in the contract. The original priced activities are normally set by reference to nationally publicised rates, e.g. BMI Building Maintenance Price Book.
- Day-work term contracts are similar to measured term contracts in so much as the contractor is reimbursed on a cost plus basis. Day-works contracts are normally only used when it is impossible to identify and schedule the maintenance activity in advance. As such, payment is made on the basis of the direct cost of labour, materials and equipment, plus a percentage for

risk and profit. One major drawback of the day-work term contract is the inherent disincentive for the contractor to work efficiently.

- Planned maintenance term contracts are normally used for the routine maintenance of plant and equipment. These contracts contain two elements: a lump sum that is paid annually for undertaking planned maintenance activities; and a schedule of rates for repairing any components which require maintenance but are not included in the original plan of work.
- Tendered schedule term contracts are similar to measured term contracts but the contract price is agreed in advance.

The BAM&R strategy should evaluate the various contractual approaches that are applicable to each maintenance and/or refurbishment activity. When evaluating alternative contract approaches it must be remembered that, whilst term contracts allow long-term working relationships to be developed and speed up the execution of maintenance activities once the contract is in place, there is no guarantee of best value for money. As a rule of thumb, the RICS guidance note on building maintenance, strategy, planning and procurement (RICS, 1999) suggests that 'project contracts should be used whenever the quality of service delivery is overwhelmingly sensitive to price fluctuations and the project value is of sufficient size to affect prices'.

Once the general approach to procuring maintenance contracts has been decided upon, the BAM&R strategy should outline how such contracts are placed and judged. The criteria should cover the following areas:

- Initial selection of contractors
- Technical competence
- Financial competence
- Performance measurements against which contractors will be judged
- Response time
- Behaviour on site (number of complaints)

In addition, with all term contracts, mechanisms should be outlined for checking that the work has actually been done and has been done to an acceptable standard. It is normal practice to only inspect a representative sample of the orders that are raised. As such, the sampling approach should be clearly outlined in the BAM&R strategy.

15.8   Maintenance and refurbishment costs

This section of the BAM&R strategy should outline the expected costs associated with maintenance and refurbishment work and describe the financing arrangements by which these costs are to be met. For most organisations this is the most contentious part of the BAM&R strategy document.

The costs of maintenance and refurbishment are normally determined through the application of life-cycle cost models. These models endeavour to predict the cost profile of maintenance and refurbishment actions over a specified time period (normally no more than five years). The models are based on the assumption that every component of the built asset has a life span. For any given component, its life span is governed by factors such as: durability, quality, climatic condition and use. Over this life span, periodic maintenance activities will ensure that the component continues to perform at an acceptable level until the obsolescence gap is such that the component is refurbished to meet the increased demands. Life-cycle costing attempts to predict the time-scale for these maintenance and refurbishment activities and then estimate the cost of them. When estimating the cost of the maintenance and refurbishment activity, due account must be taken of external factors (e.g. inflation, interest rates, etc.) that may influence the assumptions made. To this end, life-cycle cost analyses should not be considered definitive; it must be accepted that they will require periodic updating. Once established, the life-cycle cost predictions will be used to evaluate 'what if' scenarios which form part of detailed maintenance and refurbishment planning. To this end, whilst the BAM&R strategy document contains only a summary of the output from the life-cycle cost model, the detailed model should be available for maintenance planning purposes once the overall maintenance and refurbishment budgets have been set.

Whilst knowing the likely maintenance and refurbishment time–cost profile for an organisation's built asset is important, setting the budget is critical. The BAM&R strategy should outline the factors that have been considered in arriving at the budget, including the:

- Relative importance of maintenance and refurbishment costs compared to all operating costs
- Current condition of the built assets
- Maintenance history of the built assets, etc.

Furthermore, the strategy should present the outcome of these considerations, normally in the form of annual budgets covering a five-year-period, which are broken down into various subheadings.

The subheadings will depend upon the nature of the organisation and the amount of work to be undertaken.

Once the budget has been identified, the financial arrangements that are required to provide the budget must be addressed. There are many financing strategies that can be adopted and this part of the strategy is probably best done in conjunction with the organisation's finance director. It is worth remembering that no matter how well argued the case for maintenance and refurbishment is, if it does not have the support of the finance director it will stand very little chance of success when it is discussed.

15.9   Monitoring and performance evaluation

The final aspect of the BAM&R strategy outlines the procedures for monitoring all aspects of the maintenance process and reporting back to the executive board. The strategy should outline the quality performance measures that will be used including:

- Financial benchmarks, e.g. costs per square metre of buildings maintained
- Performance benchmarks, e.g. number of complaints from users
- Disruption benchmarks, e.g. down time

This section should also describe the processes by which the BAM&R strategy itself will be updated in the light of changes to the organisation's business structure and built asset portfolio.

Once the BAM&R strategy has been developed, it should be fully discussed at the most senior levels within an organisation to ensure that everyone understands the implications of the strategy. To this end, when writing the strategy documents, one must always bear in mind the audience. The strategy documents are written for, and will be used by, people who are not building experts. As such they should be written without undue recourse to technical jargon. This does not imply, however, that they are simplistic documents; on the contrary, they must present complex arguments in a clear and concise manner that informs management decision-making. Only if this is achieved will they then receive the attention, at a senior level within an organisation, that they merit.

15.10   Maintenance planning

From the day that an organisation occupies a building, its maintenance activities begin. If the BAM&R strategy has been well thought through, this need not cause any major problems. In this circumstance, the strategy will outline the approaches that should be adopted for all types of maintenance and refurbishment activity and will have identified the available budget. Thus, the challenge facing the maintenance manager is one of maintenance planning and programming.

Maintenance planning involves balancing the required maintenance and refurbishment activities with the available budget. To achieve this, it is first necessary for an organisation to explicitly establish the condition of its built assets and identify specific repair and/or refurbishment actions that are required. (Note: This process should not be confused with the life-cycle cost models that were produced for budgeting purposes. When considering detailed maintenance planning it is necessary to know what the actual condition of the built asset is, not what was predicted.) Establishing current condition, and the timescale for the various maintenance and refurbishment actions, depends upon the repair strategy that has been assigned to the various built asset components.

For components that are operated on an operation-to-failure strategy, predicting exactly when failure will occur is impossible. As such the only way to plan for such activities is to estimate the number of failures likely within a given time period (normally one year) and then set aside a budget to cover this activity. The estimate of likely failures should be based on historic records.

For components that are maintained on a programmed basis (e.g. routine maintenance of building services), the amount of maintenance necessary within the maintenance period can be assessed by reference to manufacturers' instructions. Again, these can be budgeted for on an annual basis.

For components assessed by their condition it is necessary to undertake periodic stock condition surveys to identify maintenance and/or refurbishment need. The stock condition survey involves the physical inspection of the built asset in order to establish its current condition and the costs of maintenance and refurbishment activities. Surveys are normally carried out on a cyclical basis (typically five-yearly) and record the condition of the external fabric, internal finishes and fittings, services (including plant and equipment), grounds and communal areas. Typically, for each built element (complex elements may be broken down into sub-elements), a survey would collect information on its:

- Construction type – description of the element.
- Current condition – either using a qualitative condition scale or a quantitative year to action.

- Priority of action – typically identification of any health and safety issues.
- Estimate of cost of the action – either a spot assessment or by reference to a schedule of rates.

Once all the stock condition survey information has been collated it can be combined with the operation-to-failure and programmed maintenance activities to produce a model of the likely maintenance needs, and their associated costs, over the planning period. If the estimated costs of the maintenance and refurbishment activities fall within the available budget then maintenance planning can move onto the final stage, that of packaging the work activities into suitable parcels which can then be sequenced and contracts let in line with the contractual arrangements outlined in the BAM&R strategy. Unfortunately, this situation rarely occurs in practice. It is more normal to have the situation in which the cost of the work required is greater than the available budget. In this situation the maintenance manager has to prioritise between the various maintenance and refurbishment activities. Again, this process should be simply a matter of applying the principles agreed in the BAM&R strategy.

Before leaving condition assessment it is necessary to examine the quality of the model that is produced as a result of the stock condition survey process. O'Dell (1996) examined the application of the stock condition process to English house condition surveys and concluded that, whilst the stock condition survey process gave realistic indications, in absolute terms, of the maintenance and refurbishment work required on a stock, it underestimated volume/cost of maintenance work by as much as 60% that in turn could result in a final cost estimate which could be up to a third lower than the actual cost. With such a large potential error in the cost predictions, one must seriously consider its usefulness in detailed maintenance and/or refurbishment planning. The development of an alternative methodology, probably based on quantitative measures supported by communication technologies, that can accurately predict maintenance and/or refurbishment need, is the second of the challenges facing maintenance professionals.

15.11 The challenges facing maintenance and refurbishment professionals

So far, this chapter has concentrated on presenting an overview of current best practice in built asset maintenance and refurbishment. This final section will discuss some of the emerging issues that may well force those involved in building maintenance and refurbishment to question some of this perceived best practice. The discussion is based on a number of sources, including:

- Government initiatives aimed at improving the performance of the UK construction industry (Latham, 1994; Egan, 1998; DTI, 2000).
- An increased awareness of global climate change (WCED, 1987; United Nations, 1992) and the impact that the built environment plays in the process (CIB, 1998; DETR, 2000b).
- A review of international building maintenance and refurbishment research activity (Jones & Then, 1998).

These sources, whilst being separate, are, by the very nature of the subject matter, interrelated, and as such this discussion attempts to identify issues that cut across the sources and establish the primary drivers that are influencing these issues. In formulating the discussion, two primary drivers for change have been identified: the pursuit of a sustainable built environment; and the development of new technologies and improved construction processes. The remainder of this discussion will focus on these issues.

15.12   A sustainable built environment

There is considerable interest in the subject of building sustainability. At the recent CIB world congress on sustainable development and the future of construction, 14 national reports were presented which summarised the consequences for the construction industry of adopting a sustainable development strategy in line with Agenda 21 (CIB, 1998). Whilst each country outlined their own sustainability agenda, all identified the need to consider life-cycle performance of buildings, and in particular building maintenance, as they strove to achieve a sustainable built environment.

The UK Government outlined their view of sustainable development in a national strategy paper, *Building A Better Quality of Life* (DETR, 2000b). The strategy defined sustainable development by four aims:

(1)   Social progress which recognises the needs of everyone
(2)   Effective protection of the environment
(3)   Prudent use of natural resources
(4)   Maintenance of high and stable levels of economic growth

The strategy recognised the importance of sustainable construction in addressing these aims and challenged the construction industry to provide built assets which:

- Cause minimum damage to natural and social environments
- Minimise the use of resources
- Enhance the quality of life
- Will be acceptable to future generations

During the consultation process it became clear that property and maintenance professionals had a significant role to play if sustainable construction was to be achieved. Indeed it could be argued that in the whole-life performance of the building their decisions are as critical, if not more so, than the original designer/contractor. What was not so evident from the consultation document was how maintenance and refurbishment decisions influenced the impact that the built asset has on the environment. Clarifying these issues is currently the focus of much research activity.

15.13   Sustainability and foresight

Sustainability is concerned with the impact that today's decisions have on future generations. To be able to predict these impacts it is necessary to have some idea of what the future may hold in store. To this end the UK Government established a number of Foresight Panels to try to predict what is likely to happen in the future. The various Foresight Panels cover a whole range of issues from the impact of an ageing population to retail and consumer services. Of particular interest to the present discussion are the views of the Built Environment and Transport Panel and in particular the Construction Associate Programme report, *Building our Future* (DTI, 2000). The panel comprised approximately 50 individuals drawn from business, government, science and academe who were charged with developing sets of issues and questions that they believed need to be addressed if the construction industry is to satisfy the needs of its customers, users and workforce. In their response the panel concentrated on what they believed were the 'big' issues – housing, reuse of buildings, globalisation, sustainability, IT and safety. Although maintenance and refurbishment was not one of their 'big issues', the panel's vision of the future does provide challenges to those involved in built asset maintenance and refurbishment. Some of the more significant challenges are outlined here.

15.14   Housing and construction need

Much of the built environment that will be required over the next 20 years is already in existence. However, if it is to satisfy the demands placed on it by changing, work and quality-of-life issues then it will need major modification and improvement. With more people working from home, changing household patterns and an older population generally, there will be a need for housing to provide a higher quality of life and greater flexibility of adaptable living and space requirements. Some of this improved quality of life will come as a result of the predicted 3.8 million new homes that are required in the UK over the next 20 years. The majority will have to come as a result of the modification and upgrading of the 21 million existing homes. The challenge for maintenance and refurbishment professionals is to provide the solutions to this need in a way that does not disadvantage large sections of the community.

15.15   Whole-life thinking

Whilst housing maintenance and refurbishment may only represent a small facet of the maintenance and refurbishment activity, how British industry tackles the broader issue of sustainability will affect everyone. Whilst there may be some detracting voices as to the reasons for, and potential extent of global warming, its impact can not be in doubt. Further, planning for the long-term effects of these impacts on the built environment will involve all of society.

Materials used in building generally account for some 40% of natural resource use, 30% of CO_2 emissions and 40% of waste (CIB, 1999). In addition, buildings account for about 45% of the UK's energy consumption, with approximately half of this figure being used in the buildings rather than in the appliances and processes that occur in them (DTI, 2000). Thus, one of the first challenges facing maintenance and refurbishment professionals is to improve waste and energy management, both in greater use of recycling and in improved maintenance and refurbishment practices. Indeed the report explicitly identified the importance of maintenance and refurbishment, stating that: 'One effective means of sustainability is to regenerate, refurbish, repair and maintain the UK's existing building stock for it already represents an investment in expended energy.'

To achieve this sustainable built environment the panel identified the need to apply whole-life thinking to built assets. In suggesting this approach the panel also recognises the limitations of current systems and has charged those involved in the construction industry to develop better assessment methods and sustainable indices for construction materials and components. These models must look at more than just costs. They must examine the impact of the built asset on its environment (both internal and external). Developing these models and then integrating them into the strategic maintenance management process is the second major challenge facing the maintenance and refurbishment profession.

15.16   New technologies and improved processes

New technology is seen by many as underpinning the development of new approaches to construction, including maintenance and refurbishment. Developments in material science are providing new lightweight materials and materials that have biological properties which are self-repairing. These developments will significantly increase the options available for maintenance and refurbishment. The challenge for maintenance and refurbishment professionals is to understand these new materials and to educate their clients on the whole-life benefits to their organisation.

In addition to new materials, ICT (information and communication technologies) will also drive changes to the maintenance and refurbishment process. Increased miniaturisation of computing and communication technologies coupled with advances in artificial intelligence will provide the ability to embed 'intelligent' sensors into buildings. Thus building performance will be able to be routinely monitored and results fed directly into maintenance and refurbishment needs models. Further, as e-commerce, particularly B2B (business to business) systems, become the normal means of communication, it is likely that the repair and maintenance department (whether an internal department or external organisation) will be aware of a building's maintenance needs before the occupier is. Such an approach does not sit easily within current theories of maintenance management. Indeed, this potential move to a truly 'just-in-time' maintenance model will require a significant review of the whole subject of maintenance.

15.17   Conclusion

This chapter outlines a strategic approach to maintenance and refurbishment that emphasises the need for long-term planning. The approach has evolved over the past 30 years and has been shaped by economics, politics and technology. It would be naïve to believe that we now have the definitive solution to our built asset management problems. Whilst significant steps have been taken in raising the status of maintenance and refurbishment from a back-room activity to at least the door of the boardroom, significant developments are still required if built asset management is to take its place as a strategic resource. Key amongst these developments is the response to the challenge of maintaining a sustainable built environment. This issue, like IT, will not disappear in the next 20 years. If the maintenance and refurbishment profession address it as an opportunity then it may well provide the lever that finally moves built asset management into the boardroom.

References

Alexander, K. (1997) *Facilities Management Theory and Practice*. E & FN Spon.

British Standards Institution (1966) *Glossary of Maintenance Management Terms in terotechnology*, BS 3111. BSI.

Chartered Institute of Building (1990) *Maintenance Management – A Guide to Good Practice*. CIOB.

CIB (1998) *Sustainable Development and the Future of Construction: A Comparison of Visions from Various Countries*. CIB Publication 225.

CIB (1999) *Agenda 21 on Sustainable Construction*. CIB Publication 237.

Department of the Environment, Transport and the Regions (2000a) *Construction Statistics Annual 2000 Edition*. HMSO.

Department of the Environment, Transport and the Regions (2000b) *Building a Better Quality of Life: A Strategy for more Sustainable Construction*. HMSO.

Department of Trade and Industry (2000) *Building our Future*. A consultation document of the Built Environment and Transport Foresight Panel. Department of Trade and Industry.

Egan, Sir John (1998) *Rethinking Construction* (the Egan Report). Department of the Environment, Transport and the Regions, HMSO.

Jones, K.G. & Then, S.S. (1998) State of the art review and research directions in property management and maintenance systems. Paper given at *Facilities Management and Maintenance, CIB W70* 18–20 November, Singapore, McGraw Hill.

Latham, Sir Michael (1994) *Constructing the Team* (the Latham Report). HMSO.

O'Dell, A. (1996) Taking stock: a Review of stock condition survey methods in the UK. *User-orientated and Cost Effective Management, Maintenance and Modernization of Building Facilities, CIB W70, Helsinki '96 Symposium*, 2–4 September 1996.

Quah, L.K. (1999) *Facilities Management and Maintenance – The Way Ahead into the Millennium*. McGraw Hill.

Royal Institution of Chartered Surveyors (1999) *Building Maintenance: Strategy, Planning and Procurement*. RICS Guidance Note, RICS Books.

United Nations (1992) Agenda 21 – action plan for the next century. United Nations Conference on Environment and Development.

World Commission on Environment and Development (1987) The Brundtland Report. *From our Common Future*. Report of the 1987 World Commission on Environment and Development. Oxford University Press.

Further reading

Association for Project Management (1992) *Body of Knowledge*. The Association for Project Management.

Barrett, P. (1995) Part 2, 3.4: Post-occupancy evaluation and 3.5: Data collection. In: *Facilities Management – Towards Best Practice*. Blackwell Science.

Betts, M. (1999) *Strategic Management of IT in Construction*. Blackwell Science.

Buttrick, R. (1997) *The Project Workout*. Financial Times Pitman Publishing.

Camp, R.C. (1989) *Benchmarking: The Search for Industry Best Practices that Lead to Superior Performance*. Quality Press.

Centre for Strategic Studies in Construction (1994) *The Successful Management of Design: A Handbook of Building Design Management*. CSSC, University of Reading.

Chapman, C. & Ward, S. (1997) *Project Risk Management*. J. Wiley & Sons.

Chappell, D. (1991) *Which Form of Building Contract?* Architectural Design and Technology Press.

Chartered Institute of Building (1980) *Building for Industry and Commerce: Clients Guide*. CIOB.

CIRIA 125 (1999) *Risk Management*. Construction Industry Research and Information Association.

Construction Industry Board (1997) *Constructing Success, Code of Practice for Clients of the Construction Industry*. Thomas Telford Publishing.

Construction Round Table (1995) *Thinking about Building*. The Business Round Table.

Ferry, D.J., Brandon, P.S. & Ferry, J. (1999) *Cost Planning of Buildings*, 7th edn. Blackwell Science.

Hammer, M. & Champy, J. (1993) *Re-engineering the Corporation: A Manifesto for Business Revolution*. Harper Business.

HM Treasury (1997) *Procurement Guidance, No. 2, Value for Money in Construction Procurement*. HMSO.

Institution of Civil Engineers (1998) *Risk Analysis and Management for Projects*. Thomas Telford.

Keeling, R. (2000) *Project Management, an International Perspective*. Macmillan Press.

Kelly, J. & Male, S. (2002) *Value Management of Construction Projects*. Blackwell Science.

Lawrence, D. & Pande, R. (2000) The facility audit – a user oriented design paradigm. Providing facilities solutions to business challenges – moving towards integrated resources management. Paper given at *CIB W70 Brisbane Symposium*, Queensland University of Technology, pp. 211–18.

Leibfried, K.H.J. & McNair, C.J. (1994) *Benchmarking: A Tool for Continuous Improvement*. Harper Collins.

Masterman, J.W.E. (1992) Introduction to Building Procurement Systems. E & FN Spon.

Maylor, H. (1999) *Project Management*, 2nd edn. Financial Times Pitman Publishing/McGraw-Hill.

Ornstein, S.W., Leite, B.C.C. & de Andrade, C.M. (1999) Office spaces in Sao Paulo: post-occupancy evaluation of a high technology building. *Facilities*, **17**(11), 410–22.

Riley, M. & Grimshaw, R. (2000) Post occupancy evaluation: the organisational aspect. Providing facilities solutions to business challenges – moving towards integrated resources management. Paper given at *CIB W70 Brisbane Symposium*, Queensland University of Technology, pp. 197–202.

Royal Institution of Chartered Surveyors (1997) *The Surveyor's Construction Handbook*. RICS Book Ltd.

Preiser, W.F.E. (1995) Post-occupancy evaluation: how to make buildings work better. *Facilities*, **13**(11), 18–28.

Project Management Institute (1996) *Body of Knowledge*. Project Management Institute.

Smith, N.J. (1999) *Managing Risk in Construction Projects*. Blackwell Science.

Sterlitz, Z. (1996) Unlocking potential. *Premises and Facilities Management*, September.

Thompson, P.A. & Perry, J.G. (1992) *Engineering Construction Risks*. Thomas Telford.

Walker, A. (2002) *Project Management in Construction*, 4th edn. Blackwell Science.

Web addresses

BRECSU http://www.bre.co.uk/brecsu/index.html

Building Cost Information Service, http://www.bcis.co.uk

Building Research Establishment, http://www.bre.co.uk

Construction Best Practice Programme, http://www.cbpp.org.uk/cbpp/

Construction Industry Board, http://www.ciboard.org.uk

Construction Research and Innovation Strategy Panel (CRISP) Performance Theme Group, http://www.crisp-uk.org.uk/

Department of the Environment, Transport and the Regions, http://www.dti.gov.uk/construction/index.htm

Department of the Environment, Transport and the Regions, http://www.dtlr.gov.uk

HAPM component life manual, http://www.hapm.co.uk/publications.htm

HM Treasury, home page http://www.hm-treasury.gov.uk/index.html

M4I, May 1999 (http://www.m4i.org.uk